Environmental Politics and Theory

Our current environmental crisis cannot be solved by technological innovation alone. The premise of this series is that the environmental challenges we face today are, at their root, political crises involving political values.

Growing public consciousness of the environmental crisis and its human and nonhuman impacts exemplified by the worldwide urgency and political activity associated with the consequences of climate change make it imperative to study and achieve a sustainable and socially just society.

The series collects, extends, and develops ideas from the burgeoning empirical and normative scholarship spanning many disciplines with a global perspective. It addresses the need for social change from the hegemonic, consumer capitalist society in order to realize environmental sustainability and social justice.

The series editor is Joel Jay Kassiola, professor of Political Science and dean of the College of Behavioral and Social Sciences at San Francisco State University.

China's Environmental Crisis: Domestic and Global Political Impacts and Responses
    Edited by Joel Jay Kassiola and Sujian Guo

Ecology and Revolution: Global Crisis and the Political Challenge
    By Carl Boggs

Democratic Ideals and the Politicization of Nature: The Roving Life of a Feral Citizen
    By Nick Garside

Democratic Ideals and the Politicization of Nature

The Roving Life of a Feral Citizen

Nick Garside

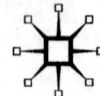

DEMOCRATIC IDEALS AND THE POLITICIZATION OF NATURE
Copyright © Nick Garside, 2013.

All rights reserved.

First published in 2013 by
PALGRAVE MACMILLAN®
in the United States—a division of St. Martin's Press LLC,
175 Fifth Avenue, New York, NY 10010.

Where this book is distributed in the UK, Europe and the rest of the world, this is by Palgrave Macmillan, a division of Macmillan Publishers Limited, registered in England, company number 785998, of Houndmills, Basingstoke, Hampshire RG21 6XS.

Palgrave Macmillan is the global academic imprint of the above companies and has companies and representatives throughout the world.

Palgrave® and Macmillan® are registered trademarks in the United States, the United Kingdom, Europe and other countries.

ISBN: 978–1–137–00865–7

Library of Congress Cataloging-in-Publication Data

Garside, Nick, 1972–
   Democratic ideals and the politicization of nature : the roving life of a feral citizen / Nick Garside.
      pages cm.—(Environmental politics and theory)
   Includes bibliographical references and index.
   ISBN 978–1–137–00865–7 (hardback)
      1. Environmental policy—United States. 2. Citizenship—Philosophy. 3. Democracy—Philosophy. I. Title.

GE180.G37 2013
363.700973—dc23   2013003213

A catalogue record of the book is available from the British Library.

Design by Newgen Imaging Systems (P) Ltd., Chennai, India.

First edition: August 2013

10 9 8 7 6 5 4 3 2 1

Contents

Series Editor's Preface vii
Joel Jay Kassiola

Acknowledgments xiii

Introduction: Democracy and the Feral Citizen 1

1 Why Wandering 13

2 Why Feral 45

3 Why Citizenship 55

4 Feral Citizenship as Method and Feral Citizen as Guide 65

5 Public Realm Theory, from State to State of Being/Becoming 85

6 A Tough Walk: Environmentalists on Democratic Terrain 105

7 The Obscured Promise of Green Citizenship 117

Conclusion: A Feral Citizen's Democratic Imperative 145

Notes 159
Bibliography 181
Index 195

Series Editor's Preface

Resuscitating Democracy, Politics, and Environmentalism through Feral Citizenship

> Wandering, for feral citizens, is not a way of getting anywhere, but a way of being somewhere...Feral citizens are content with visiting, disrupting, listening, and interrogating.
>
> —Nick Garside, Introduction, 6

It is my honor and great pleasure as series editor to introduce and welcome you to a new addition to the Environmental Politics and Theory (EPT) series. This book by Nick Garside is the third volume to be published in this wide-ranging series encompassing contemporary political and social thought in response to the environmental crisis facing our world. The series' content is evident from the subject of its two previous publications: one, a collection of analyses of the problems, governmental responses, and global implications of the dire state of the environment in China,[1] and the other, an exploration of the need and potential for revolution as a reaction to global ecological threats.[2] With Garside's contribution, the EPT series moves on to exploration of other impacts of the current environmental challenge. How should we view democratic theory—now virtually a universal value in the world, at least nominally? What is the nature of the political, in theory and political life today? And, how might the environmental movement protect nature through the lens of what the author terms and theorizes, "feral" citizenship? Garside advocates for the rejection and replacement of the dominant theory of liberal democracy, arguing that it is a nonparticipatory, self-interested, and procedurally based (as opposed to being based on substantive content) economistic perspective that pacifies citizens. While the liberal democratic theory is globally hegemonic today, Garside contends that it suppresses genuine politics and the public realm as well as commodifies society and nature (Introduction).

In opposition to the prevailing liberal reductionist and passive theory of democracy, Garside prescribes feral citizenship and democracy, and an accompanying social order that supports an authentic feral democratic life consisting of mental and physical: activity, spontaneity, and exploration or wandering. He prescribes openness to change, critical thinking, disruption of dogmatism and the belief in fixed, permanent solutions, and freedom; and, finally, rejects the reduction of democracy and politics by liberal democratic theory and society to "a procedural servant to the needs and desires of liberal economics" (Introduction).

Garside's theory of feral citizenship and democracy constitutes a fundamental challenge, what he terms a "dissent," to the current globally dominant liberal (I would term it "neoliberal") way of life based on the political value of freedom and wildness. Garside describes feral citizens as "wanderers who are also consciously disruptive actors, [they] take on the role of continuously disrupting the dangerous reduction of democratic ethics to democratic procedures" (Introduction).

To the volume's distinctive credit, Garside critically assesses liberal democratic theory and society, providing a valuable articulation of the alternative theoretical replacement of the feral citizen and a resuscitated, genuine, democratic social order. He draws upon the works of those he terms "public realm" theorists[3] who argue for an expanded—relative to liberalism—conception of politics and public space: Hannah Arendt, Jürgen Habermas, Cornelius Castoriadis, and Chantal Mouffe. He synthesizes these theorists' views to support his argument for the importance and expansion of politics and public life. In addition, Garside constructs a theory of democracy that is feral, pluralistic, and agonistic (conflictual); feral because of its constant social interrogation and criticism, pluralist because it emphasizes the rejection of absolute (monistic) knowledge as the one source of value, and agonistic because it results from individual variability, freedom, and the absence of absolute knowledge (*passim*).

Garside emphasizes the high costs to democracy, politics, and the environment of pursuing self-defeating, liberal democratic ideals and way of life. Instead, he acts as his feral citizen would by theoretically criticizing and disrupting the prevailing liberal doctrine. Garside takes the necessary further step by proclaiming the superiority of "wandering feral citizenship" (see first three chapters that carefully explicate the meaning of each of these terms by relying upon a rich and diverse set of intellectual sources, such as, chapter 1 on "wandering": Thoreau on walking, and Baudelaire and the French Situationalists' critique of modern society).

When reading Garside's depiction of the feral citizen who constantly challenges, explores, disrupts, and ultimately values a feral democratic society embracing increased public open space for theoretical and physical wandering, exploring, and disrupting, I could not help but think of Socrates as a model feral citizen.[4] Socrates, as portrayed by his student, Plato in the *Republic*, would seem to be the quintessential feral citizen of Garside's creation who constantly wanders, interrogates, explores, and eschews dogmatism or fixed, permanent views deemed invincible to criticism striving to keep the intellectual exploration process ongoing and avoid its premature cessation. For the feral citizen, as for Socrates, the ideal is to conduct endless questioning and disruption of accepted opinions by creating a social order that permits such endless inquiry within a public space.

The Socratic similarities of Garside's feral citizen may be thought to undermine his theory of feral democracy because Socrates ultimately met his demise at the hands of the Athenian elite who felt threatened by his disruptive feral nature and criticism of established beliefs. Socrates's trial culminated in a death sentence by a jury of Athenian citizens—a fate Garside understands. The feral citizen must have courage to withstand the inevitable social opposition to feral ways (the example of Socrates's devotion to endless questioning in a hostile society right up to his taking of the hemlock would seem to be quite apt for Garside's purposes).

On the contrary to qualms about the analogy of Socrates to the feral citizen, I think Socrates's martyrdom to the feral life of "visiting, disrupting, listening and interrogating" supports rather than contradicts the importance of a truly pluralist, feral, democratic social order in contrast to an individual or small minority committed to the principles and traits of feral citizenship and society as articulated by Garside. The mainstream majority always views feral citizens, such as Socrates, as threats, and will take appropriate suppressive action to silence political and social critics of the closed-minded, self-interested advocates of the status quo; hence, the importance of establishing a truly open, public realm which not only protects, but also encourages, such critical feral behavior.

It is essential to emphasize the importance of moving from the individual feral citizen to the social level of a feral democracy as illustrated by the feral Socrates and his victimization by a closed elitist society's violent backlash upon one seeking to explore and interrogate conventional social beliefs or ideas. It seems to me that Garside's presentation of the feral citizen is not intended to be a prescription for individual transformation of a single, stultified liberal democratic citizen, but rather the necessity for the society as a whole to embody the characteristics described in Garside's book.

Feral identity is not supposed to replace social movement politics, as Garside maintains in the final chapters devoted to how the theory of feral democracy could be used to assess the postcolonial, feminist, and environmental movements for societal transformation. Instead, feral theory is proposed with the goal of protecting active, substantive (not merely procedural), participatory, feral democracy even though these movements must struggle against their powerful opponents and defenders of closed-minded self-interest (chapters 4–7).

Garside's presentation of the feral citizen and expanded political space (in contrast to the devaluation and constriction of liberal politics) is intended to "incite and radicalize" (Conclusion) those who consider themselves genuine democratic victims under the limiting conditions of liberal democracy with its emaciated concepts and social institutions of commodified, instrumentalized, and unfree liberal "democracy, "politics," and "nature." However, Garside appeals to his readers to recognize that their suffering is a result of misguided social theory, and an unsustainable and alienating way of liberal democratic life. He encourages us to seize opportunities such as the environmental crisis to expand the public sphere through social movements, like the environmental one. Yet, true to his theory of wildness, Garside cautions us to be mindful of the dangers of such social action groups succumbing to the lures of fixed and permanent solutions, including liberal democracy and environmentalism, thereby threatening the very pluralist and feral democratic goals they seek to achieve.

In the epigraph to chapter 6, "A Tough Walk: Environmentalists on Democratic Terrain," Errico Malatesta, an early twentieth century Italian revolutionary, provides wise counsel and the epistemological foundation for Garside's theory of pluralist feral democracy.

> We do not boast that we possess absolute truth, on the contrary, we believe that social truth is not a fixed quality, good for all times, universally applied or determinable in advance...Our solutions always leave the door open to different and, one hopes better solutions. (1965, 269)

The last two chapters (6 and 7) of Garside's volume specifically address the environmental crisis and resulting social movement as an application of how feral citizenship and democracy can influence one's thinking. Garside critically examines prominent environmental theorists who offer various approaches to ecological citizenship (stewardship, care, and deliberation), and applies his pluralist theory of feral citizenship and democracy to the respective strengths and deficiencies of important ideas in recent environmental political theory.

In this explicitly environmental part of the volume, Garside examines how politics might change when environmental values (such as sustainability) and feral democratic citizenship are connected, and where there are barriers to such linkages. He points out how environmental political theorists implicitly (and perhaps unwittingly) adopt a "thin" procedural, liberal view of democracy with its constricted view of the political. A central idea in this portion of Garside's volume is to take the expanded and positive feral view of pluralist and agonistic politics and apply it to nature, extending politics to the nonhuman part of nature; this is where he derives the title: "Politicization of Nature."

How Garside's ideas and his theory of feral citizenship and democracy impact our thinking about nature and the environment deserve careful consideration among few developed alternatives to the liberal democratic instrumentalization of nature for human purposes dependent upon its misconceived anthropocentrism (the belief that humanity is both separate and superior to nature). Garside's unorthodox position views nature and the environmental movement through the lens of feral democratic theory where no absolute truth is acceptable. It advocates for maintaining, even improving, social conditions of open, active, participatory political thinking as crucial for achieving the desirable feral characteristics, led by freedom as described in Garside's theory of "the roving life of a feral citizen." The environmental social movement needs to advance its vital goals, Garside believes, in a manner that does not exclude humans who are manifestly a part of the environment. He quotes Hartley Dean's important insight,

> While it is perfectly true that we cannot save humanity unless we save the earth, there is no purpose in saving the earth at humanity's expense...there is no reason to reject the possibility of human emancipation. (chapter 6)

In conclusion, I would like to highlight what Garside calls a "beautiful" passage by Rebecca Solnit on the history of walking, illustrating how environmentalists might reconsider their political roles when politics is understood in a feral manner:

> Musing takes place in a kind of meadowlands of the imagination, a part of the imagination that has not yet been plowed, developed, or put to

any immediately practical use. Environmentalists are always arguing that those butterflies, those grasslands, those watershed woodlands, have an utterly necessary function in the grand scheme of things, even if they don't produce the market crop. The same is true of the meadowlands of imagination, time spent there is not work time, yet without that time the mind becomes sterile, dull, domesticated. The fight for free space—for wilderness and for public space—must be accompanied by a fight for free time to spend wandering in that space. (chapter 6)

The similarities between the ecological "meadowlands not used for a market crop" and human "meadowlands of imagination" analogized by Solnit in this passage show the value and fecundity of Garside's theory of feral citizenship and politics. The physical and cognitive wilderness, especially when expressed in public, must be protected and cherished because these human wild areas constitute what it means to be a free human, with fundamental potential for democratic, political, and environmental ideals to sustain the feral, or Socratic, traits of society.

I cannot think of more fitting symbolism for Nick Garside's inspiring and creative book about the political and environmental implications of wildness, both in nature and in our politics. Let us always defend and preserve the "meadowlands of our [political] imagination" and ecological landscape. Garside's feral theory of citizenship helps us understand and appreciate these undeveloped and free domains of land and human imagination.

JOEL JAY KASSIOLA

Acknowledgments

Many people have supported me throughout the production of this book and I could not have written it without them. My thanks go out to Scarlet Neath and Brian O'Connor from Palgrave Macmillan for helping get this book into order. I also wish to thank an anonymous reviewer who made some very helpful suggestions about an earlier draft of the book as well as Katie Haigler and Deepa John for their helpful edits during the production process. Special gratitude goes out to Cate Sandilands and Dean Bavington who have consistently pushed me on my ideas and encouraged me to keep writing, keep dreaming, and keep exploring. I would also like to thank my new colleagues at Laurier Brantford who have helped me to realize one must never stop fighting to be different. Finally, I would like to thank my family. Flora and Jack, thank you for making life the adventure it is. To my partner Elizabeth Bender, thank you for being my companion on these adventures and most importantly thank you for being my source of emotional and intellectual support during production of this book and more importantly in life.

INTRODUCTION

DEMOCRACY AND THE FERAL CITIZEN

In the opening paragraph of *On Revolution*, Hannah Arendt (1963, 1) argues that "no cause is left but the most ancient of all, the one, in fact, that from the beginning of our history has determined the very existence of politics, the cause of freedom versus tyranny." Feral citizenship in general, and this book in particular, is a response to the way the cause of freedom has been recast within a political climate where "democratic governance has achieved the status of being taken to be generally right" (Sen 1999, 5) and where "liberation discourse has moved from socialism to democracy" (Fotopoulos 2001, 5).[1]

As democratic governance is taken to be generally right, much of the present-day activity around revitalizing and repoliticizing public space is developing relative to a supposition that, along with the grounding of liberation discourse within democratic theory, victory in the political battle Arendt speaks of may soon be coming to the forces of freedom. Yet, as I show in this book, assuming the forces of freedom have won the classic battle is premature. While there is a recognizable shared allegiance to democracy, the condition has not, in most cases, led to a revitalization of public space, a challenge to institutionalized authority, or a sufficient rethinking of social relations. More typically, the lip service to democracy has led to a shallow endorsement of a sort of *mass* or representative democracy, in which citizens are assumed to be passive or are pacified, and politics as active public debate and engagement with others is replaced by what might be better described as a common allegiance to procedural statecraft.

Rather than advancing the forces of freedom, the collective endorsement of democracy (at least by those granted the voice to endorse) may well have come at the cost of the political ethos that makes democracy something for which to struggle. On this note, Michael Saward (2003, 4) has explained that "to evoke democracy can at the same time be to attempt to fix a [favorable] meaning to it." Fixed meaning, of course, is never favorable to everyone. Indeed, if democracy is viewed as a commitment to the ethical ideals originated in the French Revolution, as Chantal Mouffe, Ernesto Laclau, Claude Lefort, and Alexis de Tocqueville suggest it ought to be, or if it is considered emergent from the early Greek polis, as Cornelius Castoriadis, Hannah Arendt, Murray Bookchin, and Takis Fotopoulos suggest it ought to be, the thin commitment to present-day procedural-deliberative forms of decision making represents as much of a threat to the continuation of the democratic tradition as it does an opportunity to further the necessary and permanent challenge to authority implied by a substantive commitment to democracy.

Nearly 30 years ago, Benjamin Barber (1984, 24) suggested we (those of us living in liberal democratic societies) have reached the point where freedom has become "indistinguishable from selfishness and is corrupted from within by apathy, alienation, and anomie; equality is reduced to market exchangeability and divorced from its necessary familial and social contexts; happiness is measured by material gratification to the detriment of spirit." If Barber is right, democracy, as both an ethic and a practice, has been reduced to little more than a procedural servant of the needs and desires of liberal economics.

However, just as it would be premature to assume the forces of freedom have won the ancient battle, it would be equally premature to hand the victory over to the forces of tyranny. Actually existing democratic regimes are once again being challenged by citizens who face very real problems that can no longer be controlled by the largely unreal spectacle of democracy that current representative forms of democracy rely on. Perhaps most importantly, there are numerous theories of the public realm that continue to offer helpful lenses through which to explore the promises and limits of the democratically inspired political moments erupting all over the world as we begin the twenty-first century. In fact, Jürgen Habermas, Chantal Mouffe, Cornelius Castoriadis, and Hannah Arendt, among others, each suggest radical and substantive ways of participating in a society in which democracy has attained default status.

This book introduces feral citizenship as a way of adding to the forces of freedom. As a method of political engagement, feral citizenship represents a way to traverse the terrain that democratic culture has created. As wanderers who are also consciously disruptive actors, feral citizens take on the role of continuously disrupting the dangerous reductions of democratic ethics to democratic procedures. The reduction of substantive ethics to procedural norms is a reduction Castoriadis (1997b, 1) believes could only originate in "a crisis of the imaginary significations that concern the ultimate goals [*finalités*] of collective life," and something, I believe, could only happen in a democratic society free from active and engaged citizens.

Along with Castoriadis, I see democratic regimes as tragic regimes. I also see democratic creation as "the creation of unlimited interrogation in all domains, what is true what is the false, what is the just and the unjust, what is the good and what is the evil, what is the beautiful and the ugly" (1997a, 343). Along with Mouffe, I see modern liberal democracy as essentially paradoxical as "the condition of possibility of a pluralist democracy is at the same time the condition of impossibility of its perfect implementation" (2000, 16). I agree that when it comes to understanding the uniqueness of the kind of democracy that has been established in the West over the last two centuries the primary influence is the "dissolution of the markers of certainty." Until this indeterminacy is viewed as foundational to democratic society, the liberatory possibilities of the current pluralist democratic conditions and "proliferation of particulars" will never be recognized or realized. I am, though, less sure of the value of the democratic and liberal articulation than Mouffe, and believe the democratic tradition has far more to offer than even a *radically democratic* liberal democracy can promise. Thus, the favorable definition of democracy that I defend is not immune to Saward's critique that those who define democracy tend to fix particular and favorable meanings to their definitions.

The project undertaken in this book is located in an analysis of politics that stresses its contentious, critical, and conflictual nature. Yet, while accepting the irreducibility of tension and antagonism, I also defend what might be perceived as a particularly pure vision of politics for the purpose of clearly distinguishing political acts and citizenship activities from more typical, dominating, and instrumental acts. On the one hand, to keep the distinction I, like Mouffe (1992b, 238), believe that within modern pluralist times the citizen-individual distinction along with the private-public (individual-*res publica*) distinction, can be maintained without sustaining a corresponding belief in discrete separate spheres. On the other hand, aligning myself more with republican notions of political agency as an opportunity to be free, I distance myself from Mouffe by insisting that the tension between liberty and equality is not one between my freedom as an individual and my duties as a citizen, but one between my opportunity as an active democratic citizen and my struggle to remain free from any fixed position that hinders my identity as a rambling, disruptive, feral citizen.

The fact that some kind of commitment to democracy is an essential component of any contemporary political theory is not lost on environmentalists. In fact, over the past decade environmentalists of all stripes have struggled to reconcile the need for environmental sustainability with the desire for democratic freedom. This ongoing struggle makes green political thought an ideal space for the feral citizen to wander into and ask questions of. The last third of the book takes on this task by examining what happens to politics when the link between environmental desires and democratic citizenship is made. The political potential of the partnership is immense but so too are the hindrances that must be addressed before the liberatory promise of this articulation is realized. A feral cituizens' travels into environmental terrain is adventurous, largely antagonistic, and full or uncertainty; what more could a political wanderer want?

It is not a mere coincidence that Hannah Arendt and Chantal Mouffe, the two most explicit celebrants of plurality and agonism among the above mentioned public realm theorists, emphasize the need for *traveling* individuals throughout their work. Arendt, for example, is convinced of the need for visiting and traveling in order to attain more truthful[2] representative thinking, and when discussing the human condition of action she argues that

> *trespassing* is an everyday occurrence which is in the very nature of action's constant establishment of new relationships within a web of relations, and it needs forgiving, dismissing, in order to make it possible for life to go on by constantly releasing men from what they have done unknowingly. Only through this constant mutual release from what they do can men remain free agents, only by constant willingness to change their minds and start again can they be trusted with so great a power as that to begin something new. (1958, 216, emphasis added)

Forgiveness is needed for the actor to be free to be more than s/he is, to leave the temporary foundation s/he may have constructed, and ramble on in order to continue visiting and trespassing. For Arendt, it is by traveling and visiting that more relations can develop and the previously unthought can emerge.

A slightly different traveling metaphor is found in Ernesto Laclau and Chantal Mouffe's early work where they suggest that within the condition of modernity,[3]

> the *terrain* has been created which makes possible a new extension of egalitarian equivalences, and thereby the expansion of the democratic revolution in new directions. It is in this *terrain* that there have arisen those new forms of political identity which, in recent debates, have frequently been grouped under the name of "new social movements."
> (Laclau and Mouffe 1985, 158, emphases added)

They then, shortly afterward, make the point that

> the democratic revolution is simply the terrain upon which there operates a logic of displacement supported by an egalitarian imaginary, but...it does not predetermine the direction in which this imaginary will operate...The discursive compass of the democratic revolution opens the way for political logics as diverse as right-wing populism and totalitarianism on the one hand, and a radical democracy on the other.
> (Laclau and Mouffe 1985, 168)[4]

The references to terrain made by Laclau and Mouffe suggest that while it may be true that democracy is threatened on many fronts, there is still a plethora of liberatory potentials that can be unleashed with a better understanding of, and approach to, the conditions of pluralist democracy. Arendt's reference to trespassing and visiting likewise suggests that the opportunity for political engagement remains even within modern liberal times. The above mentioned attempted partnership between environmentalism and democratic theory is a prime case in point. What I find strange, however, is that these traveling metaphors are not attached to a theory of citizenship. Furthermore, as prevalent as references to terrain, trespassing, and visiting may be, and as essential as engagement with others is to democratic practice and theory, there is scant theoretical discussion concerning how to ensure the terrain is protected and maintained. One of the most unfortunate consequences of this absence is a lack of dialogue around the relationship between active citizenship and traveling. In this book, I suggest a perpetually amateur citizen—one who is free from political affiliations, unfixed, and indeterminate—is the kind of political wanderer that concurrently needs and can help to focus attention on democratic terrain.

In her fabulous book *Wanderlust*, Rebecca Solnit (2000, 72) explains that "part of what makes roads, trails, and paths so unique as built structures is that they cannot be perceived as a whole all at once by a sedentary onlooker. They unfold in time as one travels along them, just as a story does as one listens or reads, and a hairpin turn is like a plot twist, a steep ascent a building of suspense to the view at the summit, a fork in the road an introduction of a new storyline, arrival the end of the story." The only way to perceive the path is to experience it, to travel along it, and to participate in the story it tells. Democracy is just such a path, but democracy is never a built or complete structure. Rather, it is an always-expanding path with endless offshoots, side trails, and ways of traversing. Anthony Arblaster has described democracy as

> not only a contestable concept, but also a "critical" concept that is a norm or ideal by which reality is tested and found wanting. There will always be some further extension or growth of democracy to be undertaken. That is not to say that a perfect democracy is in the end attainable, any more than is perfect freedom or perfect justice. It is rather that the ideal is always likely to function as a corrective to complacency rather than a prop to it. (1987, 6)

Thus, democratic terrain is constantly expanding and being rediscovered as citizens risk the political and take on the role as corrective to complacency.

As with any trail, different travelers enjoy certain sections more than others, and some areas are easier to access and more traveled than others. Likewise, some paths that were once well traveled are now overgrown and difficult to negotiate, with offshoots that are almost impossible to find. Yet, the trail and all its offshoots remain and wandering along it, rambling past its real and imagined boundaries and divisions, and revisiting and creating new offshoots is what active citizens do.

Feral citizenship offers a unique and pleasurable way of experiencing and traveling along the path, creating new paths, wandering off the path, keeping the path open, and ignoring the signposts that may have been erected by those maintaining the path. Wandering, for feral citizens, is not a way of getting anywhere, but a way of being somewhere. Wandering offers a particular perspective that eschews expertise or responsibility while allowing for spontaneity and curious adventure. When attached to citizenship it allows for genuinely political exploration; when attached to a consciously feral identity it allows for disruptive introduction of political moments. Pleasurable wandering may be a privilege, but it is also something that most everyone is capable of doing, wants to do, and ought to have the opportunity to do. Whether wandering is physical or theoretical, it offers a unique perspective on one's surrounding, a perspective absent from far too many political journeys.

As a perpetual wanderer, the feral citizen engages, visits, and learns from others. As s/he is also free from a commitment to any particular theory, s/he can actively listen rather than translate, co-opt, or attempt to convince those s/he meets. However, as a consciously feral visitor, the feral citizen inevitably disturbs and incites communities s/he visits. Ideally, this disturbance will politicize the communities and stimulate democratically informed self-reflection. As an eternal nomad, the feral citizen will be absent from the rebuilding stage. Rebuilding is a task left to those who have been disturbed but still wish to retain their home(s). As a curious and committed political agent who requires communities and houses to visit, a feral citizen is not a nihilistic subject intent on destroying that with which s/he comes in contact. In order to perform the political act of disruption, feral citizens will always need communities to visit, interrogate, and disturb. As such along with not having permanent political affiliation (beyond a commitment to democracy), the feral citizen is likewise not interested in creating a movement or following. Feral citizens are content with visiting, disrupting, listening, and interrogating.

Michel Foucault once said,

> I dream of the intellectual who destroys evidence and generalities, the one who, in the inertias and constraints of the present time, locates and marks the weak points, the openings, the lines of force, who is incessantly on the move, doesn't know exactly where he is heading nor what he will think tomorrow for he is too attentive to the present; who wherever he moves, contributes to posing the question of knowing whether the revolution is worth the trouble, and what kind (I mean, what revolution and what trouble), it being understood that the question can be answered only by those who are willing to risk their lives to bring it about. (1988, 124)

The approach to citizenship I propose in this book, while more explicitly political and hopefully less exclusionary than Foucault's "intellectual" (citizens do not need to be intellectuals), has much in common with Foucault's dream. The primary difference, as my use of the term "citizenship" implies, lies in my desire to link this type of subjectivity and methodology with amateur political agency. While I acknowledge the shallowness of the present-day common allegiance to democracy, I also find a great deal of promise in the fact that "democratic governance has now achieved the status of being taken to be generally right" (Sen 1999, 5). I am therefore unwilling to concede the concept of democracy to mainstream reductionist uses of the term. Democracy needs to be reclaimed and resituated so it can once again unleash its disruptive and critical promise. It is less "this is what democracy looks like" than it is "this is why democracy should still scare you!"

Overview of the Book

This book is made up of this introductory chapter, seven main chapters, and a conclusion. In chapters 1, 2, and 3, I explain the relevance behind the three key terms—"wandering," "feral," and "citizenship"—I use to describe the approach to democratic citizenship that I propose. While these are separate chapters, an understanding of the relevance of each term is directly related to understanding the rationale behind the other two terms; so ideally, they should be read together in order to appreciate the particular attributes of wandering feral citizenship as a whole. Starting in chapter 1 with wandering, I give a historical account of figures and groups who recognize the value of wandering. Included in this group are Henry David Thoreau, George Sand, Mary Austin's Walking Woman, British ramblers, Parisian flâneurs and Situationists, and various other advocates of the aimless wander.

After defending the democratic value of physical and mental wandering, I offer a warning by bringing in a number of "uses" for walking that while not necessarily apolitical nevertheless threaten the political promise of wandering, including the pilgrimage and the march. In this section of the chapter, I draw parallels between pilgrimages and social movements and defend the position that the wandering and distinctly political characteristics of citizenship need to be kept separate from the more instrumental features of the pilgrimage and the march. I argue that the democratically induced proliferation of particular interests, while politically important is also a potential threat to the broad democratic political ethos as too much of a focus on movement toward an achievable goal could replace the nondirectional and intrinsically valuable wandering attributes of politics with the marching or pilgrimage attributes of a movement ethos.

My intent is not to criticize social movements or to substitute a social movement ethos with a political ethos; it is rather to ensure that social movements retain space for and value the less instrumental aspects of their progressive intents.

I defend and celebrate all kinds of walking and see political worth in all forms of walking (except the kind done on a treadmill). Presently, the sort of peripatetic activity most under threat is wandering, and the kind of movement that has the most parallels to politics is wandering, thus it is wandering that I spend most of my time defending.

After defending the pleasure and democratic relevance of wandering in chapter 1, I then go on to chapter 2 to explain the relevance of claiming the disruptive term "feral" as a descriptive part of amateur political agency. In this chapter, while I do give a brief historical account of certain societal responses to feral children and animals, I concentrate on describing how a feral identity can allow consciously feral citizens to become disruptive agents of change in modern day democracies.

In chapter 3, I defend the value of, and need for, active citizenship of all sorts. Democracy has always relied on citizens who participate in the public sphere and act on their freedom to decide their own future and destiny. I argue that a democratic citizen who accepts the privilege to wander and uses this privilege to disrupt, challenge, and learn from those met along the democratic terrain offers a unique, radically democratic way of celebrating the political, expanding the political, and embodying the promise of democratic culture.

I conclude chapter 3 by bringing together the three metaphors to explain how the spaces created by wandering feral citizens produce informal micropolitical moments that offer all those present in the moment the opportunity to act as spectators, storytellers, and actors—the three identities Hannah Arendt correctly describes as necessary for the realization of political moments. Feral citizens are not the only creators of such moments but through acts of citizenship and direct engagement with those encountered they offer a particularly playful yet equally political way of traversing the democratic terrain and embodying the promise of a politics-first society.

Chapter 4 explains that due to the general acceptance of democratic ethics as a necessary foundation of a just society, contemporary democratic terrain is littered with potential sites for the wandering feral citizen to visit, learn from, and disrupt. One of the most apparent and welcome developments along the terrain has been the creation of "we" spaces among individuals who are not only demanding to be recognized but also through the creation of these "we" spaces are performing as desiring agents who insist on being considered on their own terms. Many of the demanding groups and associations have politicized the shadowy realms of the private sphere and unearthed the hidden oppressions of putatively inclusive deliberative procedures that continue to be used to legitimize actually existing democracy. For wandering feral citizens, the interest in these "we" spaces lies in their attempt to house difference and antagonism within spaces that retain the capacity to bring people together as a collective. In this chapter, I show that there are key political insights to be found in the way difference has been transformed from a problem to deal with to a central feature to embrace within certain postcolonial and feminist "we" spaces. The chapter ends with an explanation of how wandering feral citizenship can transfer lessons learned from feminist and postcolonial struggles into a defense of an expanded political sphere that continues to encourage the articulation and disarticulation of "we" spaces within a pluralist democratic society.

Following this explanation into how feral citizenship can learn from engagement with "we" spaces, I return, in chapter 5, to theories of the public realm to show a few trails feral citizens would do well to travel upon. As many of the theorists I examine in this chapter continue to influence my particular approach to feral citizenship, I return to them as what Chantal Mouffe (1993) might call a friendly enemy, guided by a strong commitment to increase the opportunity for democratic tension and plurality. Public realm theorists, through their own participation in the development of political theory, create particular public spheres. They also, due to their democratically inspired interest in the public sphere, offer their own interpretations and strategies for the (re)vitalization of the public sphere. My specific concern is with how the *particular theories* of public realm theorists such as Jürgen Habermas, Hannah Arendt, Cornelius Castoriadis, and Chantal Mouffe can help nurture the broader democratic requirement for tension-filled public space and discourse. The plurality of agents and theories struggling over occupation of the public sphere frustrates those who desire simplification, control, order, and predictability. Taking up the methodology of feral citizenship, I question and critically engage with the strategy of each theorist in an attempt to uncover the promise and the danger (the threat to democratic ideals) of each strategy in relation to the substantive view of democracy required for feral citizenship. I have no interest in simplification.

In chapters 6 and 7, I look at environmental politics in order to explore the liberatory potential of the body of theory developing around green citizenship. Green citizenship represents what I believe to be a promising, yet once again politically dangerous, turn of events within the democratically promising green public sphere.

When environmental issues first entered public arenas of debate and discussion, many progressives believed environmentally oriented movements and theories contained the most radical implications and critiques of the nation-state, of capitalism, and of parliamentarism (e.g., Bookchin 1994 1). However, like so many other initially revolutionary movements, the disruptive implications of these radical roots have never been realized. Green public spheres have yet to draw adequate attention to any genuinely radical alternatives to the status quo. Nevertheless, I believe the turn to green citizenship, once the implied commitment to democracy is taken seriously, offers a way of reinvigorating the disruptive, critical, and radical promise of environmentalism.

Chapters 6 and 7 show that the politicization of nature offers disturbing and explicitly democratic potential that ought to be recognized and celebrated by those who wish to expand the public sphere and defend the priority of the political. Unfortunately, it also shows that green political theory is subject to the same struggles that tend to limit the political promise of most movement-inspired politics. While theorists of green citizenship show clear intent to ground green politics in democratic discourse, there is an understandable longing to use the legitimacy of democracy and citizenship as a means to attaining predetermined green desires.

As nonhuman nature cannot adequately participate in political discourse and we cannot entirely "know" nature, there is a deep challenge to democracy and environmentalism implicit in the turn to democracy within green political thought. It is impossible and undesirable even to try to represent, know, or manage a true nature. The political struggle is to find a way to discuss and talk about the relation with nature outside the confines of the desire or need to know. The reason green citizenship has not led in this direction lies in the tendency to ask citizenship to do too much. It is with the creation of new political moments within the green political sphere that such problems can be brought to light. As with the previous chapter, many of the theorists I examine in these two final chapters have inspired my own theory of feral citizenship; so once again, I return as a friendly enemy who wishes to disrupt but certainly not destroy the community.

There is no intent, in this book, to replace movement politics with wandering politics. The intent is to use the method of feral citizenship to help remind those involved in movements of their reliance on democratic culture. By consistently focusing on the need for democratic ethics and the revitalization of the public sphere, the acting feral citizen creates opportunities for engaged political discussion that focuses not on an achievable end point but on a need to protect the terrain that allows such struggles to continue, and contribute to, something much grander such as the ongoing battle between the forces of freedom and the forces of tyranny. There will always be more to be done, more to discuss, and more to debate. This is politics.

CHAPTER 1

WHY WANDERING

In the next three chapters, my intention is to tap the metaphorical richness of the image conjured up by the idea of the wandering feral citizen. To assist in this task, I explain the inspiration behind the choice of the three particular terms central to the idea. My rationale is simple, wandering equals aimless movement, feral equals disturbance, and citizenship equals primacy of the political. Individually, each term carries particular inferences that help clarify and guide the characteristics and activities that inform feral citizenship and free feral citizens. Collectively, they represent a radically democratic approach to political agency that focuses first and foremost on the intrinsic value of democratic politics.

Wandering, for example, is a figurative—and at times, real—practice of feral citizenship inspired by wanderers who embodied, recognized, and defended the privilege of aimlessness and noninstrumental exploration. Wanderers have particular political relevance as they have often defended the pleasure that accompanies their existence, at least for a time, outside or in temporary ignorance of the ever-worsening conditions of their respective societies. While rarely intentionally political, I show that wanderers represent important figures with innumerable implications for radical democratic citizenship that become most apparent once politics, like wandering, is itself acknowledged as a terrain for pleasurable exploration and expression.

My particular focus in this chapter is on defending and clarifying the need for the distinctiveness of wandering in the face of the present-day entanglement of new social movements, which is making it particularly challenging for progressives to speak of freedom in times of clear and present need. I do this not by claiming the superiority of wandering over other socially necessary activities, but by defending aimless movement as a particularly important component of political agency needed for the realization of a healthy and critically engaged democratic culture.

To better situate the discussion of wandering in present-day pluralist conditions, I bookend the chapter by drawing parallels between the reduction of walking to treadmill training, and the reduction of democracy to procedural decision making. After discussing the joy of wandering I suggest that new social movements, as important and essential as they are to pluralist democracy, are often more akin to pilgrimages than they are to wandering and thus, threaten to replace politics with more instrumental excursions in which spontaneity and adventure are replaced by end-focused and (at least partially) self-interested actions. In each case, most of the activity occurs within the confines of what I, following Guy Debord (1988) in *Comments on the Society of the Spectacle*, refer to as a typical dynamic (state/economy/media) that excludes the amateur, inefficient, unpredictable, and pleasurable attributes of wandering and active politics.

Amateurism and the Treadmill

It has often been said that *ideas need legs*—a prophetic statement if there ever were one—yet with no enticing places to walk, legs get weak and walking becomes more a burden than a pleasure. Ideas settle and representatives of those ideas become comfortable with their own interests and beliefs. In such a society, the political terrain may still be used but its presence is more assumed than celebrated. Gradually, creativity, joy, and imagination associated with political life are replaced with more immediate needs of particular and often parochial ideas, interests, or resistances.

In relation to walking, the most extreme example of present-day reduction has to be the treadmill.[1] This machine, once used as punishment for prisoners, is now threatening to replace the once joyful and socially interactive act of wandering with the safe, monotonous, individual, and rhythmically efficient burning of as many calories as possible in the shortest time period. The mere notion of treadmills alters the way walking as an activity is discussed, and exemplifies the prevalence and imposition of notions of efficiency and speed. A similar threat with equally detrimental outcomes can be said to face democracy and politics. The presence of "representatives" and voting booths, along with the broad reduction of politics to decision making and struggles with the state for rights, alters the way citizens think about political involvement and effectively reduce politics to a tool for extrapolitical desires. Treadmills replace potentially wonderful and risky aspects of walking with the "result" of better health—with safe, repetitious, uninterrupted, and quick burning of calories. Locating politics within procedural decision-making bodies and largely administrative spheres similarly represents the replacement of potentially joyful interactive engagement with more inclusive, measurable, and thus presumably more legitimate, decision making bodies.

Treadmills and procedural decision making correspondingly take one minor component of their respective activities and make it appear to represent the whole; once unintended *outcomes* become the stimulus, purpose, and sole reason for taking up the activity. In both cases, it is the least enjoyable, most disciplinary, and dutiful aspect of the activity that is emphasized. In politics, inefficiency, amateurism, and exploration are gradually squeezed out, replaced by trite duties that cannot help but deter participation and discourage broad involvement.

Such a reduction of politics to a means rather than an end in itself is a cautionary tale that celebrants of the proliferation of particular movements should not ignore. Resistance to this threat lies in the realization of a common enemy and the yet to be fully appreciated political promise of the abundance of social and occupy movements that draws attention to the common enemy at the same time as it creates the links between the previously isolated and typically issue focused movements. These links also help to create enticing new terrain that can encourage those with radically democratic and antiauthoritarian commitments and ideas to take a stroll. Furthermore, the vastness of the actions and the spontaneity of the performances mean most of the places have yet to be controlled and are not enticed by the spectacle. Audience becomes each other and those present in the moment, and common enemies begin to emerge not as any particular corporation or political party but as the foundation and norms that allow such parties and corporations to act as they do.

A second equally germane parallel between the treadmill and decision making appears when the story of the history of the treadmill comes to light. "The original treadmill," explains Solnit (2000, 260), "was a large wheel with sprockets that serve[d] as steps that several prisoners trod for set periods…Their bodily exertion was something used to power grain mills or other machinery, but it was the exertion, not the production, that was the point of the treadmill." Solnit, quoting James Hardie's 1823 book on the treadmill, continues, "It is the *monotonous steadiness* and not its severity, which *constitutes its terror,* and frequently breaks down the obstinate spirit" (2000, 260, emphases added). It does not take an especially imaginative leap to see how the citizen as a passive voter faces the same spirit-draining monotony. Indeed, as numerous studies of "political participation" have shown, the tedium of voting along with the recognition that the political system like the treadmill never changes, never goes anywhere, and always does the same thing regardless of who is on it, has led to a not surprising decrease in participation in elections and authorized political activity. However, as all else appears to be lost there is a consistent mass outcry of injustice whenever examples of lost opportunities to vote or vote tampering are discovered. One always wants to choose, but remarkably most fail to see that the choices have been made for us long before we get to take on our legitimizing role of making the final decision. Don't vote and the choice is theirs; vote and the choice is theirs. The argument will be further clarified later in the book but here we can state that *decision making is to politics what the treadmill is to walking*.

Henry David Thoreau's essay "Walking" is perhaps the most explicit defense of the particularity and promise of the kind of wandering that could never be reduced to activities on a treadmill. In this lovely short piece, he clearly understands and celebrates the promise of aimless wandering:

> We should go forth on the shortest walk, perchance, in the spirit of undying adventure, never to return—prepared to send back our embalmed hearts only as relics of our desolate kingdoms. If you are ready to leave father and mother, and brother and sister, and wife and child and friends, and never see them again—if you have paid your debts, and made your will then you are ready for a walk. (1993, 108)

Thoreau found few "who understood the art of walking, that is, of taking walks—who had a genius, so to speak, for sauntering." For Thoreau, the particularity and privilege of walking without purpose or necessity is what allows for creativity. The act itself is an art, a pleasure lost once any need or other purpose intrudes on the wander. Similar to performance art, yet without the audience or sphere of appearance so important to making performance political, the activity itself, rather than the produced outcome, is the purpose. To saunter or roam relies on the capacity to free oneself from that which keeps one from leisurely pleasures, for when one can walk "there will be so much the more air and sunshine in our thoughts" (Thoreau 1993, 109), and so much the more opportunity to be guided by desire and spontaneous urge.

For Thoreau, "life consists with wildness. The most alive is the wildest. All good things are wild and free [and] in Wildness is the preservation of the World."[2] Therefore, unstructured and undisciplined activity, thought and being may not be needed to survive but it is needed if one is to live. Wildness is something beyond our knowledge, it can never be controlled or domesticated, it is where the previously unconsidered may emerge, and it is where spontaneity is most prevalent. Like Thoreau, I "rejoice that horses and steers have to be broken before they can be made the slaves of men, and that men themselves have some wild oats still left to sow before they become submissive members of society" (1993, 114). A wild and curious wandering citizen set free from the controlling and imposing norms and institutions of liberal democratic society seems to me to be precisely what democratic politics needs. It also resists the push to be broken.

A consistent wanderer initiates, discovers, and uncovers the new, but this newness occurs primarily (solely, if we agree with Arendt) when the movement is not instrumentally guided. Arendt, like Thoreau, knew that the ability to find pleasure in aimlessness was not something necessarily given to all. The opportunity needs to be available to everyone who wishes access to political space, but like the opportunity to wander, not everyone is required to participate. What feral citizens can do is create the conditions that will entice more and more people to venture along paths that have no obvious end. In addition, like Thoreau's equation of wandering with art, Arendt (1958) sees politics as a sort of performative art practiced by those who are free from the realm of necessity where survival, not life, is the focus. Arendt's celebratory and playful approach to politics will be discussed in much greater detail in chapter 5. Here, what is important to note is, first, her obvious support for wandering and the noninstrumental movement that accompanies the wanderer and, second, her recognition that not everyone will participate in politics in this manner. Wandering is one particular approach to traveling the democratic terrain; it remains politically significant only if there are places to wander to and others who are *not* wandering. A political wanderer's activities involve physically and theoretically visiting and exploring sedentary communities along the political landscape; rather than enticing others to give up on their communities and join the walk, the intent is merely to engage these others as a political wanderer committed to democratic ethics and the expansion of the public sphere.

The equation of politics, art, wandering, and pleasure are particularly relevant to feral citizenship.

The Ramblers

Attempts to limit spaces for pleasurable walks have been responsible for politicizing walkers for many years. This is nowhere more evident than with the populist Ramblers in Great Britain. According to the Ramblers' Association's official web site (http://www.ramblers.org.uk/), the roots of the organization go back to the nineteenth century, when a growing number of British residents turned to the countryside for rest and relaxation in the face of expanding industrialization. Quite unlike the North American approach to wilderness protection, British Ramblers, who were largely from the working class,[3] focused on ensuring right to access and freedom to roam in the countryside including mountains, moors, heathland, downland, and registered common land under the Crown ward. While North America was protecting and incarcerating its wilderness (Birch 1995), British workers were demanding access to their countryside and rambling into areas that were beginning to be enclosed and designated private by wealthy landowners buying up the land. Unlike Thoreau's solitary wanderer, Ramblers tend to travel in groups and when purposefully political, organized mass trespasses that intentionally drew attention to their illegal activities.

Ramblers were politicized because they and others who shared their passion for walking the countryside were being denied access to spaces to which they believed citizens had absolute rights.[4] The simple act of rambling became illegal once fences were erected and private property signs were affixed where previously paths were traveled. Many ignored the signs and rambled on but others wanted to draw attention to the loss of public space and the enclosure of the commons. For those who wished to draw attention to the enclosures, the battles were and are less over a demand for increased rights than a demand to acknowledge the rights of the public as well as the right to access public land. Ramblers have always fought for the right to roam, and by holding "Forbidden Britain," mass trespasses used acts of civil disobedience to (re)open the British countryside for those who wish to wander through public land. Without the actions of the Ramblers, previous crown land that was gradually being bought up by wealthy British landowners would have been off-limits to the public.

As recently as 1997, the Ramblers' Association was successful in convincing the British Labour party to support and legislate the "right to roam," a law that allows access to right-of-way paths and open countryside across Britain. The law came into effect September 19, 2004, and while consistently threatened nevertheless represents an inspiring assault on gated privilege. And "after a long mapping process the new right of access to mapped areas...was fully implemented on Monday 31st of October 2005" (http://www.ramblers.org.uk/freedom/righttoroam/history).

The achievement of the right to roam along with the famous rights-of-way battles at Kinder Scout[5] and other places across the British countryside are ideal examples of politically relevant walking acts and stories.[6] Wanderers who would not accept predetermined boundaries and borders politicized their walking by actively trespassing. The whole idea of not allowing public access to the countryside regardless of who owned it was made a political issue by Ramblers who were unwilling to adapt passively to the changing conditions.

Feral citizens can learn plenty from Ramblers as they tend to be guided by their own imperative, have traditionally ignored borders and boundaries erected on what they perceive to be public land, and have been driven by a passion to protect opportunities they believe should not be limited to the privileged few. Most Ramblers are not self-described radicals, only passionate walkers who as a result of their passion demand access and the opportunity to ramble associated with the realization of such a demand. While they do get laws changed, their most relevant actions were, and still are, oriented toward conscious denial of rules that restrict common access. The mass actions have meant the authorities have little choice, either change the laws or accept the fact that many will and do openly ignore them.

The relationship to the state seems to be as much one of pay attention to us as it is one of demand for justice. Laws are secondary or responsive to the actions not primary or natal and as such, the relationship between the state and citizens becomes altered as the state is forced—if it wishes to retain legitimacy—into a responsive role and its authoritative position becomes questionable.

There is another important trait common to Ramblers and feral citizens. Whenever explicitly political activities are the point of a walk, not all Ramblers are expected to participate, but all those who love to ramble and walk the countryside benefit from the actions of the participants. Likewise, not all democratic citizens are required to be feral citizens or practice feral citizenship, but ideally all active citizens benefit from the actions of those who do. The trespassing actions increase the sphere of political engagement by opening up opportunities to learn from others, by increasing the number of discursive spaces among those who desire the democratic ideals of freedom and equality, and by uncovering opportunities for political articulation. The trespassing that feral citizens must do involves venturing into communities that populate the political terrain but only rarely or indirectly help to invigorate and extend the political sphere's domain. The trespassing of feral citizens is more symbolic than actual, but the desire to gain access to and republicize space along the political terrain has numerous parallels to the Ramblers' actions. Celebrants of rural wandering and the numerous Ramblers organizations help to show the political relevance of wandering as an aimless art form practiced by travelers who may follow legitimate trails, trespass on forbidden trails, and may also, at times, struggle for the right to roam wherever they please. All these traits, along with an opportunity to be momentarily free from the burdens associated with dealing with life's necessities, have important political consequences when embraced as characteristics of citizenship. Yet, as relevant as these wandering acts and actions are, when it comes to *wandering as an artistic way of life* and a stimulus to noninstrumental politics, influences on feral citizenship are as much urban as they are rural.

Urban Strolling: Bathing in the Crowds

In the *Oxford English Dictionary (OED)*, pedestrian as an adjective is defined as dull and uninspired. To do something pedestrian is to do something as a commoner with little originality and even less inspiration. Parallels to the modern-day citizen as little more than an occasional yet responsible voter are apparent as are the more welcome links to common opportunity "to do something as a commoner" but like the citizenly acts common pedestrian acts can lead to uncommon, artistic, and unpredictable experiences that cannot be satiated as long as the opportunity to wander remains.

Rebecca Solnit (2000, 213), referring to de Certeau's chapter on urban walking in *The Practice of Everyday Life*, suggests that "if the city is a language spoken by walkers, then a post-pedestrian city not only has fallen silent but risks becoming a dead language, one whose colloquial phrases, jokes, and curses will vanish, even if its formal grammar survives." For de Certeau (1984, 97), "the act of walking is to the urban system what the speech act is to language or to the statement uttered." Walkers give the city its life and like discursive functions that constitute both subjects and objects, the act of walking in the city perpetually recreates the cityscape as a platform to be written upon and experienced uniquely by wanderers, cruisers, loiterers, and travelers of all types.

Unfortunately, yet not unexpectedly, it is not just the keepers of the safe city with their criminalization of loitering and distrust of pedestrians that is threatening to render free movement obsolete. Actually existing democracy is itself becoming a postpolitical administrative system making the language and ideas that support the defense of politics for its own sake as rare and obtuse as the language of the city. Indeed, like wandering, it is nearly impossible to defend the intrinsic relevance of politics within the confines of the global language of commerce and instrumentalism. Wanderers of the city and wanderers of political terrain are threatened species in much the same way as paths to wander along and wild spaces to explore are threatened spaces.

Wildness, creativity, and genuine individuality have always worried those who are creating, maintaining, and conserving society's norms. Politics and wandering are always a little mysterious; they offer opportunities for spontaneity and surprise and as activities they can never be fully contained by rules and order. They offer opportunities to break free from the monotony of survival and allow one to dream, imagine, and potentially perform a better or at least other world. Given their unpredictability and disruptive nature, it is not surprising to find both activities (wandering and noninstrumental creative political engagement) beginning to be framed as unwelcome, invasive, frivolous, and a general nuisance by the reasonable authority of those maintaining the urban and political terrain.

In essence, democracy and walking in the city have been successfully reduced to usually burdensome and banal acts that need to be endured by those unfortunate enough not to have access to better means to solve the perceived issue at hand. The treadmill and the voting trough have allowed walking and acts of citizenship to become subject to judgment tools and perceptions that make it ever more difficult to defend the particular value of the activities. One can burn calories more efficiently and safely on a treadmill than by walking; one can move from A to B faster by relying on any one of many available alternatives to walking; one can reach decisions on important public issues far more efficiently if elections, plebiscites, and referendums are not required. All of these statements are true and without the language of the city and the wandering adventures of the celebrants of politics, we may lose the ability to point out the obvious fact that these statements are responding to questions and problems many of us never asked or care to know about. Fortunately, thanks to the absence of any need for exclusionary skill or the commonality of both practices, the language and activity of wandering and politics are not yet dead.

As hinted at above, my primary interest is in peripatetic aimlessness as a political trait that encourages a way of being and thinking that informs a type of activity especially relevant to democratic citizenship. Paris has produced two such characters or species of wanderer that are the subject of the next few sections, the flâneur and the situationist. While the latter has been much more politically involved and culturally influential,[7] the former was more explicitly focused on wandering and fantasy. Thus, each will be shown to be particularly significant to feral citizenship, if for different reasons.

Wandering the City: The Flâneur

The characteristics and actions of the original Parisian flâneur of the nineteenth century were first, and I believe best, described by Charles Baudelaire (1863) who was also, not coincidentally, the first to formally articulate the term "modernism."[8] He wrote numerous poems and offers many examples of flânerie but his lyrical prose work "The Painter of Modern Life" represents a definitive defense of modernity or the present one finds oneself in and thus offers important insight into what made the flâneur a marginal and heroic figure of his own modern life as well as an important influence on the modern-day feral citizen. The poem itself is fascinating on numerous fronts as it can be read as a representation of a particular kind of modern art, a description of a particular flâneur as artist, an example of a written manifestation of flânerie itself, and, more generally, a celebration and defense of the passion for life in a city with spaces for adventure. Here we look primarily at the poem as a description of a flâneur by a fellow flâneur.

The particular flâneur or artist who is the subject of "The Painter of Modern Life" is Baudelaire's friend Constantin Guys, but it can be assumed that most contemporary readers would not have known this fact and if they did, they likely would not know who Guys was. Guys, who wished his works to be discussed "only as though they were the works of an anonymous person" (Baudelaire 1863, 1), never signed his paintings,[9] refused to be categorized as *merely* an artist, and consistently resisted public recognition. In fact, he only reluctantly allowed Baudelaire to write about him on the condition his name not be included in the poem. We know this because the prose itself begins with such a note, which his original readers obviously would not have seen. But leaving out Guys's name was more than merely granting a request. The request and the absence are both indicative of the flâneur's particular method and abhorrence of glory. Modesty and anonymity take on specific relevance for the flâneur who wants to be *in the crowd but not of the crowd*, for to be recognized is to lose one's identity as a flâneur who merely wants to be left to live one's own observational or voyeuristic life.

Baudelaire explains that Guys was an artist and a man of the world, a forever-curious cosmopolitan traveler who was subject to wanderlust.[10] His art was an expression of his character sustained by his passion for movement, love of modernity, and the adventures of the everyday. As Baudelaire (1863, 2) explains, "he takes an interest in everything the world over, he wants to know, understand, assess everything that happens on the surface of our spheroid." Curiosity about the wonders that are hidden in the city and the denizens of the city drives the flâneur's passion and feeds his art. He is "at once a dreamer, a historian, and a modern artist, someone who transforms his observations into texts and images" (Gleber 1999, viii).

As an incessant observer of his environment, the flâneur is constantly looking for "that indefinable something we may be allowed to call 'modernity'" (Baudelaire 1863, 4), but seems to intuitively know better than most that the opportunity to engage in the search is the true promise of modernity. Recognizing the need for unpredictable and wonderful places to travel, the flâneur was different from the pedestrian who "would let himself be jostled by the crowd," the flâneur "demanded elbow room and was unwilling to forgo the life of a gentleman of leisure" (Benjamin 1968, 12).[11] He wanted to be left to be who he was, a man with room to roam as well as enough leisure time to take advantage of that room.

Flâneurs were typically lonesome characters so they did not entice others to follow; they enjoyed their stylistic otherness during the day and expressed it during the night and wanted little more than to keep it that way. Even when the realm of necessity did impose itself on their life and they had to sell their wares (poems, pictures, stories) in the market place, the relationship was such that the selling was usually viewed as a necessary evil rather than a determining goal.[12] If art was to be sold, it was simply to allow the flâneur to remain a flâneur.

It is true that flâneurs were not essentially political characters, yet it is also true that if Constantin Guys is representative, the flâneur is much more than an idler struggling for a slower pace of life. Reducing the flâneur to a passive spectator of modernity is as unfair as reducing him to a mere stationary artist. Importantly, as this aspect seems to get lost in certain modern discussions, a flâneur's gift of seeing and hearing is connected to his gift or passion for expressing himself.[13] The passionate nature of the flâneur could never be contained solely in a spectator's curiosity. Baudelaire clearly shows that the flâneur's wild passion extends to the way the gift of vision is expressed:

> Whilst others are sleeping, this man is leaning over his table, his steady gaze on a sheet of paper, exactly the same gaze as he directed just now at the things about him, brandishing the pencil, his pen, his brush, splashing water from the glass ceiling, wiping his pen on his shirt, hurried, vigorous, active, as though he was afraid the images might escape him, quarrelsome though alone, and driving himself relentlessly on. (Baudelaire 1863, 4)

This kind of zealous expression shows that Baudelaire's flâneur (Guys) was much more an active and influential part of Parisian society than he is typically given credit for.[14] Stories told by flâneurs are unique and continue to give insight into parts of Parisian life that would never find their way into mainstream texts. As dwellers of and in the city and celebrants and observers of the present, they had a taste for the city and a way of expressing that taste that was artistic first and informative second. It lacked the pressure of an audience or judge and can best be understood as a hybrid public-private expression of a hybrid individual whose stories, poems, pictures, and expressions were necessary excretions from a life of flânerie.

In "The Painter of Modern Life," Baudelaire saved his most critical comment for painters of his time who chose subjects of a general nature yet "insist[ed] on dressing them up in the fashion of the Middle Ages, of the Renaissance, or of the East." "This is," he argued, "evidently sheer laziness; for it is much more convenient to state roundly that everything is hopelessly ugly in the dress of a period than to apply oneself to the task of extracting the mysterious beauty that may be hidden there, however small or light it may be" (1863, 5). This lovely defense of *the modern as subject* of art is a suggestive comment for those wishing to resituate certain traits of the flâneur or flânerie in our own *modern* times. Would we not be committing the same laziness if we merely transported the flâneur of the past into the present? Wandering and politics, I have already suggested, ought to be considered art forms. The role of feral citizenship is to situate and perform the art as politics in modern times.

The resituating I am proposing is not only in time and place; but it is also in activity. I want the aimless and leisurely traits of flânerie to become a recognized and accepted part of a particular type of political agency. Interesting theoretical and physical terrain does exist throughout the public sphere, it is just hard to find.[15] Unlike the largely apolitical Parisian flâneur, a modern-day political flâneur cannot only adapt to the available terrain, but s/he must also purposely help to create it. Like the flâneur, an acting feral citizen celebrates the present, has no interest in attaining followers, does not try to convince others to embrace her/his methodology, and primarily wants to be allowed to wander political terrain as a marginal and mysterious figure. S/he may be tolerated but rarely, if ever, revered.

Feral citizens have no nostalgic view of a pure and wonderful past; what they have is a passionate commitment to reinvigorate and celebrate the present by extending the political sphere and reinvigorating political debate and performance. There are, however, vital differences between flânerie and feral citizenship that should not be overlooked. Feral citizens want to protect their spaces to wander but so too do they want to disturb them, and while they do not necessarily want everyone to become feral citizens or practice feral citizenship, they do want the opportunity for others, if they so choose, to meander through the wilderness of democratic terrain.[16] They see themselves as having far more agency than the flâneur yet they also realize that political subjectivity or citizenly action is temporary and feral citizenship is one possible way of acting on the momentary opportunity. One is never only and always a feral citizen. These particularities make feral citizenship a far less dominant part of one's life than it seems flânerie was a part of the flâneur's life.

Political flânerie as a way of dwelling in and acting on democratic terrain can offer a particularly broad understanding of an irreducible need for plurality, difference, and tension within the public sphere. Lacking expertise or a strong commitment to any particular cause (other than democracy), the political flâneur can perform the essential roles of disturbance and critique; however, if such activity is going to be democratic it requires that flânerie be available, at least on a temporary basis, to everyone.[17] It also would be helped if flânerie became a part of an explicitly political identity that celebrates amateurism and the democratic condition.[18] It is thus the idea and methodology of flânerie, not the flâneur himself, that is most readily transferable to the modern political sphere.

The flâneur's peripatetic passion brought on by a sense of emptiness and a belief that satisfaction could be anywhere—which meant little more than it was not *here* (Tester 1994, 10)—is the same passion that keeps democracy active and necessary.[19] The flâneur was always wandering, searching for satisfaction or fulfillment because wherever he was, he was neither satisfied nor fulfilled. Feral citizenship is inspired by the same lack of comfort. Feral citizens know a democracy is never something one has; it is something one strives toward; in the striving is where democracy and politics is realized. Never at home, because as a political agent s/he is constitutionally homeless, a feral citizen is a wanderer led by passion and desire; feral citizens find and at times create paths, streets, and spaces to "hang out" on the political terrain.[20]

The political focus of feral citizenship is particularly important if the typical aim and disciplinary interest of the modern flâneur is indeed, as Bauman, Buck-Morss, and Benjamin all agree, consumption.[21] According to these modern theorists, the flâneur's marginality has already been co-opted and marketed by department stores, the demands of a consumer society, and capitalist mass consumption.[22] The desire and passion that once drove the originality of the flâneur, we are told, is now sold to crowds of empty shells who are little more than consuming automatons, believing they can purchase individuality and creativity for themselves.

The threat of commodification is certainly a real one, but I refuse to accept that it is all encompassing. As rare as moments of modern nonconsuming flânerie are (or the moments when one acts outside this prescribed consumerist flânerie), such moments and acts remain. My claim is that the observational and active methods of flânerie do have something to offer modern citizenship by way of understanding, exploring, and acting upon the potentially liberatory spaces within modern pluralist society. As those spaces close down, feral citizens can create new ones and through their acts remind the masses that one's freedom and individuality does not lie in consumptive acts.

Must Wandering Be Gendered?

Before we further consider the potential role for the politicized flâneur of the present, we should note that all the wanderers and flâneurs included up to this point have been men.[23] Thus, our other modern question must be: Is flânerie and wandering essentially a male act or an act done only by men?

In answering this question, it would be difficult to argue against the claim that the "unabashed and unadulterated pleasure in the sights, views, and images of the street seems reserved for the experience of male spectators" (Gleber 1999, 171),[24] or that the flâneur "could pick and choose where to play his game—to the flâneuse [female flâneur], most of the flâneur's favorite haunts were out of bounds" (Bauman 1994, 146). It would also be difficult to deny the fact that the freedom female flâneurs gained through access to consumables and general sexual liberation "had the nightmare effect of 'freeing' all women to be sexual objects (not subjects)" (Buck-Morss 1986, 124).[25] However, like most rules there were exceptions and the exceptions are far more interesting than the norm. Nevertheless, when it comes to the norm I do not go as far as Janet Wolff (1985) in arguing that there was and is no such thing as a flâneuse. Rather, I am in agreement with those who suggest that it is most likely that female flâneurs were simply not noticed by male history writers as the case of George Sands discussed below attests to. I do, however, accept the claim that for the flâneuse as flâneuse rather than disguised flâneur what may have been the end of flânerie for the male flâneur (arcades replaced by artificially lit and climate controlled department stores in a rationalized society) was the beginning of a new kind of flânerie for the female flâneur. The shift to a more consumerist society does threaten the promise of flânerie but it does not eliminate it. The increase in shops and cafes as well as the extended hours for consumption made the streets safer for women and created opportunities for women to be in the crowd but not of the crowd. Furthermore, if the male flâneur is to be forgiven for selling his wares in order to remain a flâneur surely the flâneuse can be given the same liberty to occasionally consume in order to not be labeled a prostitute.

In this brief discussion of the flâneuse, there is the emergence of a slippage from unique to normalized that I do not wish to endorse. Window shoppers in an adventureless city that has turned the exploring spaces of flâneuristic adventure into the boring monotony of a consuming highway do not represent the promise of the future. Fuller streets, more cafés, and a better nightlife are all welcome ways cities can become more inclusive and safe but if there are no adventures to be had in these cities, the promise of flânerie is indeed lost. Walter Benjamin is convinced the days of the flâneur are a thing of the past; whether this is true is up for debate but not in the pages of this book. My interest is in the attributes and actions of the unusual and distinct characters who roam the city, and if we are to think of the flâneur as an example of such an actor and if we are to consider flânerie as a worthwhile influence on feral citizenship, it would be an injustice not to include the heroic efforts of unique female flâneurs not mentioned by Baudelaire or Benjamin.[26]

George Sand,[27] perhaps the most famous of the female flâneurs, for example, "entered the world as a female flâneur in disguise, functionally outfitted for that purpose in male clothes, in pants and boots" (Gleber 1999, 173). As far back as the early nineteenth century, Sand pioneered strategic boundary crossing and trespassing. Sand refused the conditions imposed on her as a woman and used whatever means available to experience the freedom, among other things, of not being subject to the male gaze. By refusing to conform to what she was supposed to be, she was able to experience a world only open to those who were male. The resistance undoubtedly helped contribute to her work as a pioneer for women's rights as well as an influential critic of elitism and class society.

In the nineteenth century, Paris women could not be flâneurs as women, but creative and marginal figures like Sand could practice flânerie. Participation required drastic disguise and a complete burying of her *visible* identity, which while challenging nevertheless gave her the opportunity to achieve complete anonymity that was particularly useful in her case as she also needed to hide her class affiliation.[28] Sand trespassed, ignored, and by participating in activities only considered suitable for males, actively denied the phallogocentric limits on her liberty, and thankfully, as there is very little written on or by female flâneurs she took the time to write to others of her illicit experiences.[29] Forced to always perform and hide her "female" identity, Sand's particular experience may be representative of a more complete form of flânerie as she was more unique, more marginal, and more politicized than other already voluntarily marginal flâneurs. In the classical flâneur sense, she was both hidden and seen and most certainly in the crowd without being a part of the crowd. Like other flâneurs, she was a contradictory figure that bathed in the masses but wanted elbow room; she not only embraced the liveliness of the city but also demanded solitude so as not to be discovered; and she liked seclusion while concurrently needing the masses as sites in which to be a stranger, an observer, and a dweller. Like Guys, she was passionate and expressed herself on paper but beyond the rest of the flâneurs she was also a cross dresser who performed as a male in public as well as when writing. She had a very clear difference between her public and private identity.

The pleasure and freedom of wandering experienced by Sand may have been similar to those experienced by men, but the story that emerges from the wandering is distinctly different as she had to resist far more societal norms, her presence was always deviant, and she was genuinely disguised. In fact, while Sand's wandering deeds may have been hidden her public words written in books suggest Sand's influence was more akin to the right-of-way battles of the Ramblers than to the wanderings of her anonymous flâneur counterparts. She believed access and ability to wander ought to be universally available, and by publishing under the pseudonym George Sands was able to clandestinely use an equally exclusionary platform to share her reflections on the experience she had as a temporary wanderer of the Parisian arcades.

Wandering, to Sand, was liberatory; her struggles for greater access and more opportunity to wander are politically important struggles that suggest wandering can be liberatory for everyone. Her efforts also suggest the opportunity to ramble ought not to be exclusionary, and if made more available and safe for women, could lead to many unexpected and previously unconsidered adventures and performances for women. As the above indicates, Sand was a far more explicitly political agent of change and resistance than were her flâneur counterparts.

The experience of Sand's female flânerie was truly *wonderful* and necessarily short lived.

> With those little iron-shod heels, I was solid on the pavement. I flew from one end of Paris to another. It seemed to me that I could go around the world. And then, my clothes feared nothing. I ran out in any kind of weather, I came home at every sort of hour, I sat in the pit at the theatre. *No one paid attention to me*, and no one guessed my disguise… No one knew me, *no one looked at me*, no one found fault in me; I was an atom lost in that immense crowd. (Sand quoted in Gleber 1999, 173, emphases original)

In the above quote, Sand alludes to certain opportunities open to women not subject to the ridiculous fashions and norms of the time. She experienced a time when she was free from society's disciplinary control and reveled in it. Unlike her male counterparts, she needed to be free not only from the realm of necessity and oppression but also from the related male gaze. Her invisibility was much more difficult to achieve and required a great deal of conscious effort and performance. *Like* her male counterpart, while being a flâneur, she merely wanted to be invisible and enjoy the pleasures of seeing without being seen. But she was always in grave danger and *unlike* her male counterpart, if she had been seen, she would have been considered much more sinister as she was blatantly deceiving all others in order to experience and perform the forbidden; her mere presence was a challenge to societal norms.

If Sand is a great example of an urban flâneuse that at least disturbs the male-centered vision of the flâneur, then Mary Austin's "The Walking Woman" represents a nice contrast to Thoreau's male solitary wanderer in the woods. Austin's "The Walking Woman" is a semifictional short story of an independent and willful woman on a solitary and endless stroll through the California's San Joaquin Valley. Like Sand, Austin uses the freedom associated with movement, roaming, and dwelling in order to offer critical insight into the social realities women of the time faced. She also, again like Sand, hints at what might be gained by stepping out of the societies' expectations and disciplinary impositions.

We learn early in the short story that the walking woman began her walk as a result of an illness brought on by her duty to care for an invalid. How long ago that was, we are not entirely sure but we do know the confinement of the care room made her sick and the open desert became the place of healing. The full story of the walking woman's past is kept from the reader until the narrator meets her for an extended amount of time in a space of isolation and leisure.

Prior to the meeting, the narrator of the tale hears many stories of the legendary walking woman known to the locals as Mrs. Walker. The often-fanciful stories turn her into an almost mythical traveler who is typically both the subject in and the subject of numerous stories and events. As interesting as the stories are they rarely offer insight into who the walking woman was as she is far too complicated a person to understand in a quick encounter. What they do indicate is that those who had the privilege of an encounter are changed and/or surprised by their interactions with her and they want to tell of their encounters. In fact, a great part of her mystery and intrigue lies in the fact that nearly all the stories told about her offer more insight into the way Mrs. Walker impacted the storyteller than they offer an accurate glance into the walking woman's way of life. The focus on the events and encounters rather than on the woman herself, we can assume, would please the walking women as she "talk[s], never of herself, but of things she had known and seen" (Austin 1907, 216). Indeed, in meeting the walking woman and attempting to tell the story of the walking woman, Austin herself has been given the gift of surprise and change. Mrs. Walker's story may come across as a harsh or sad story but she has lived and she looks fondly on the opportunities that have come to her as a result of the chance to be free and experience a fulfilled life. When the narrator does get the chance to learn a little more about the walking woman's life, she uncovers a few of the events that have helped turn the walking woman into the protagonist of her own story.

The most significant event for the walking woman occurred in a sand storm she found herself caught in, shaped by, but not beaten by. In what Lowe (2000, 27) describes as a sort of combined surrender to the forces of nature as well as a resistance to the forces of nature (and the social creation of women as incapable of resistance to such forces), the walking woman "discovers her abilities as an individual rather than a lady who must be protected." The event took place in the spring on the south slope of Tehachapi where the walking woman settled with a shepherd after a day of traveling, "pricked full of intimations of a storm" (Austin 1907, 218). The shepherd went by the name of Filon Geraud and happened to be all alone due to his companion shepherd leaving for a three-day trip during a susceptible time of year "when there is a soft bloom on the days, but the nights are cowering cold and the lambs tender, not yet flockwise" (Austin 1907, 218).

A storm hit in the morning after the walking woman arrived and by mid afternoon the flock had broken leaving Filon and Mrs. Walker to deal with the situation. In recounting the story to Austin, the walking woman exclaimed, "Until that time I had not known how strong I was, nor how good it is to run when running is worthwhile." With great effort and apparent untiring commitment, the two were able to bring the flock back together and hold the flock long enough to outlast the storm. After sharing the story and taking a moment to reflect, the walking woman began to speak about how this event changed her,

> "For you see," said she, "I worked with a man, without excusing, without any burden on me of looking or seeming. Not fiddling as women work, and hoping it will all turn out for the best. It was not for Filon to ask, Can you or will you. He said, Do, and I did. And my work was good. We held the flock. And that," said the Walking Woman, the twist coming in her face again, "is one of the things that make you able to do without others." (Austin 1907, 218)

It is also, of course, one of the things that helps make those moments when she was with others that much more rewarding for the fortunate interlocutor. The walking woman stayed with Filon until the fall, and it was within this time frame the walking woman had and subsequently lost the three things she considered most essential for a fulfilled life—to work with another, to love another, and to have a child. While the walking woman always ends up alone, all of her essentials are about relationships with others. Yet, the walking woman's return to solitary life is not about regret or lament. It is about appreciating and experiencing the moment not for its means to something else or its potential permanency but for its value there and then. As Lowe (2000, 27) explains, "the story rejects mating and reproduction as a women's sole role, for the baby dies, the shepherd moves off with his flock and the walking woman remains alone in the desert." Even after losing what she sees as the keys to happiness, she carries on her journey seemingly content with the gift of once having had that which few ever attain.

What can a modern-day celebrant of the wander intent on reinvigorating the public sphere and reprioritizing the political learn from the story of this most unique character? To start off, the feral citizen could gain an appreciation for short stories in the realist tradition but whether this is of direct relevance to feral citizenship is a discussion for another time. There are a number of more direct links that feral citizenship has with the walking woman.

The walking woman is freed from society's conventions not through economic privilege but through a denial or dismissal of the conventions that turn a person into a subject of societal norms. The walking woman loses her name, loses her home, loses her role as a caregiver, has very few material possessions, and even loses the three things she cares most about, work, love, and child. Yet, each of these losses is less lamented by both the narrator and the walking woman than it is seen as an opportunity to travel and be free. The walking woman gains the opportunity and will to take adventures not open to those with responsibilities attached to what she once had but now has lost, or perhaps more appropriately has transgressed and left behind.

The lifestyle of the walking woman is a challenge to domestication, to the norm of settling down, and to the idea of ownership or chattel but it is far from glorious, and while clearly a protofeminist story it is not a call for others to follow her particular ways. It is, however, a cry for resistance and a call to recognize the promise and pleasure in the unusual, in individuality, and in the capacity for everyone to do the unpredictable.

In chapter 5 of *The Human Condition* Hannah Arendt (1958, 157) writes, "With word and deed we insert ourselves into the human world, and this insertion is like a second birth, in which we confirm and take upon ourselves the naked fact of our physical appearance." To have this second birth and insert ourselves through word and deed we need to rely upon, yet also be free from, the conditions that constituted our first birth. The walking woman had a second birth. She inserted herself into the world by freeing herself from the conditions attached to her first birth, which defined and shaped what she was or was supposed to be due to convention and societal assumptions. A second birth is a rare occurrence but it is a key purpose of a democracy to create the conditions where such a second birth can occur.

The walking woman did not focus on challenging enforced physical and social restrictions typically imposed on women. What she did was she ignored them and then acted, traveled, and used her newly discovered opportunity to achieve self-realization and freedom. It was her individuality and willfulness that ensured she not see herself as a sole to pity but as a woman with opportunities not given to many others. The walking woman was homeless, aimless, embodied, and a denizen of the desert. As a traveler with no direct place to go, she was able to relate to her surroundings and accept the terrain as a part of her story. The terrain was less used than it was experienced. Mrs. Walker never mentions walking as a part of the happy life but there seems little doubt that it would always remain a central feature of who Mrs. Walker was.

The walking woman also contains a refutation of the care ethic as the core role for woman. This is important for two reasons. The first is that human capacity should never be contained within a core role and the second is that the care ethic is not free from societal pressures and thus should never be embraced without very careful consideration of its essentialist and limiting traits. To be sure, Mary Austin does see nurturing as a natural female trait, and her depiction of both woman and nature as subject to masculine society is based on an essentialist view of woman and nature. She never, however, allows women to be reduced to this essentialist trait and freeing oneself from this imposed (whether social or natural) role is clearly a liberatory act in Austin's stories.

The exceptional cases of Sand and Mrs. Walker do not answer the question of whether wandering as a way of knowing and interacting is an essentially male act, or whether it is an act done by males. They do, however, show that wandering has been and can be taken up by exceptional women. If we consider these cases from the perspective of inclusivity and freedom, the problem, I believe, is not that movement and wandering are male attributes; it is that men (white, heterosexual men from the leisure class) have traditionally been the only ones able to move, to wander, and to roam.[30] Unfortunately, this has meant that even those women who have taken up the activity have done so as disguised men but while this past is unacceptable it has also allowed a few women to offer exceptional stories that are full of insight.

What the above-mentioned experiences suggest to me is that expanding or democratizing the opportunity to roam is a worthy and not necessarily exclusionary goal for committed democratic citizens. I am leery of the parallels to the idea(l) of inclusivity associated with similar attempts to let others in to a previously male world and it is not enough to simply state that more should have the opportunity to wander. What I am not leery of is the implications attached to the intention for inclusivity of opportunity. If indeed more should have the occasion to wander, those agreeing to such a statement have the obligation to make more diverse avenues open. Hindrances must be explored and brought to light and a good way to do that is to investigate, to trespass, and to push the boundaries of what counts as acceptable entrance into the political sphere.

So is the political sphere a haven that can eliminate gender exclusivity? Is the political public sphere defined and organized around democratic ethics the inclusive space women have been rightfully demanding? No, it is not. However, those acting within political public spheres cannot justifiably deny the need to attempt consistently to free the political sphere from prejudice of any sort; they cannot legitimately consider such questions solved and they cannot, if they are at all true to their democratic commitments, ignore the legitimacy of demands for equality that go beyond mere equal access to predetermined political procedures, duties, and responsibilities. As Amartya Sen (1999) argues in relation to democracy, those who would deny equal opportunity, access, and the right to stimulate discussion must make a convincing argument against the request; what such an argument would entail is hard to imagine.

An extension of politics is necessary, but it alone is not enough to ensure equality on any level. What counts as politically relevant must consistently be revisited and expanded; structural and ideological hindrances to participation must be regularly acknowledged and altered and the way we define and respect what is political must consistently be rethought and challenged. As all these needs are a central part of democratic ethics and culture, active and engaged wandering citizens will perform these tasks.

When considering gender the extension of politics and democratic justice requires *seeing the political realm not as gender neutral but as offering conditions for equal opportunity for all to wander in and on their own terms.* The space must be organized around the ideal of the equality of unequals where equality is itself recognized as a contestable, unachievable, and evolving ideal. Before this ideal can be realized and stimulated by feral citizenship the question of how to approach the conditions of modern society must be better addressed. Figurative practices of wandering are entirely different from past peripatetic acts when considered in the context of the capitalist-Western world's society of the spectacle I describe in the next section.

Seeing the City: Debord's Society of the Spectacle and the Situationists[31]

If French philosopher-situationist Guy Debord is correct, the flâneur's imaginative creation of a fantasy world has been embraced, normalized, and institutionalized first in order to help construct the fetishism of the commodity[32] and then, more recently, the fetishism of the image. If we are living in such a society of the spectacle, the dividing line between fantasy and reality has, according to Debord, become almost entirely indistinguishable. Or, to borrow the language of the flâneur, the real has been elbowed out of all available mediums in order to be replaced by a fantasy world that is then fed back to the masses through the unavoidable media that has masterfully repackaged the flâneur's imaginative ability to live in a fantasy world free from the impositions of reality.[33]

In a society of the spectacle, the formerly unique and creative attributes of the flâneur are so universalized and homogenized by the media, economy, and state dynamic that even the creators of this integrated spectacle have lost sight of the difference between the fantasy and the actual world.[34] The practitioners have preached the gospel so often and have seen the success of the fantasy so clearly that they too can no longer find the dividing line between real and fantasy. The spectacle encompasses and controls the acts of the producers of the fantasy as much if not more than the consumers. The privileged producers may not see themselves as part of the crowd but they are undoubtedly both in and of the spectacle.

CEOs of large corporations, rebels, and compliant spectators are equally scripted in the play created, directed, and perpetually repeated by the media/economy/state apparatus (what Debord calls the typical dynamic within the integrated spectacle). The society of the spectacle appears to be the inevitable result of bureaucratic or administrative pseudodemocracy where no one takes or has responsibility for governing but everyone is still governed. Reminiscent of tyrannical society where unaccountable arbitrary rule was the norm, this new condition may actually be more insidious as there is no tyrannical government to focus one's attention on, there is no tyrant controlling the tyranny, and most are unaware of their status as subjects of tyrannical control.

The brilliance, endurance, and horror of spectacular society lies in its ability to distribute disinformation[35] and allow enough choice and create enough situations to keep individuals believing they are actually individual and free. A society of the spectacle has no center, no goal, and is interested almost solely in self-preservation. It offers space for critique and praise but determines the content. Indeed, there are still paths to walk along but each is a prescribed and controlled path, which may appear as particular, different, and even unique but is typically part of predetermined trail systems with paths to satiate or break the thinking individual.

"The spectacle," argues Debord (1994, 20), "preserves unconsciousness as practical changes in the conditions of existence proceed. The spectacle is self-generated, and it makes up its own rules: it is a specious form of the sacred." As it is the institutionalization and protection of a *particular* fantasy world, the society of the spectacle has much more power than the flâneur ever did. The spectacle represents one fantasy and it is so all encompassing and authoritative that all other fantasies are housed in and created by it. It is also much less liberatory as not even the beneficiaries of the society of the spectacle are aware, let alone in control of, the fantasy.

If the integrated spectacle is all encompassing, there is no chance for critical and disruptive democracy. Thus, I believe it is far too simple to assume the spectacular relationship of the integrated spectacle is the only power dynamic that guides contemporary democratic subjects. If this were true, there would be no Guy Debords and certainly no feral citizens. The masses may well be primarily made up of passive spectators, but rarely if ever have the masses (as a group) been the source of original ideas. It has always been the unusual or unique few that disrupt and challenge the sheepish masses.[36] Once more people become aware that resistance to a particular injustice or falsehood, as necessary[37] as it may be, is not resistance to the spectacle, and once more of the real does reappear as is happening around the world with the spectacles attempt to respond to the so-called economic bubble with austerity and blind allegiance to old ways, spectators will find their agency. Resistance to particular injustices is essential in making the spectacle *appear* more real but so too is it essential to making the real appear. A great deal of the "realness" of resistance depends on its appearance in the spectacle, which is precisely the unreal, but many acts of resistance ignore the spectacle or are ignored by the spectacle. Unlike the spectator whose "own gestures are no longer his own, but rather those of someone else who represents them to him" (Debord 1994, 23),[38] the modern actor is a performer who lives in the moment and cares little for the reality of the spectacle.

So while Debord's description of spectacular society may be accurate on many fronts, it is premature to accept that independent thought and action are nostalgic dreams of prespectacle times.[39] Even pacified spectators can be rattled out of their role as mere observers as the spectacle crumbles and gaps in the never-complete fantasy are made evident. Social movement advocates have helped draw attention to the gaps at the same time as they may have helped to make the spectacle appear more real. I am thus not suggesting that social movement advocates and others who operate within the spectacle permanently abandon their "necessary" role as part of the society of the spectacle. In fact, as dangerous and limited as any action within the spectacle may be, as long as there are remnants of the spectacle there will remain a need to use the spectacle for one's own purposes, and such use does not always render one complacent or a victim. The trick is to find ways not *only* of being part of the spectacle, but also when part of the spectacle, being conscious of the game being played. To realize the limits of the spectacle it is important to play other games where the spectacle is less dominant; difficult as they are to see and participate in, such places do exist.[40]

Despite his cynicism, Debord, at least before he wrote *Comments on Society of the Spectacle*, was himself a fellow wanderer and celebrant of real life as part of a group called the Situationist International. When describing the beliefs of the Situationists, Debord (in Knabb 1995, 17) explains, "First of all we think the world must be changed. We want the most liberating change of the society and life in which we find ourselves confined. We know that this change is possible through appropriate actions." To help initiate such changes, Debord and the Situationists proposed creating temporary situations that brought dead space to life. He wanted to jar people out of their automaton life and remind them of their opportunity/capacity to think and live rather than merely to behave and survive. Situationists were intent on creating spaces where individuals could interact together as people not mediated by commodities. What the Situationist did was create *real* situations.

Given that they called themselves Situationists, it is worth noting just what a "situation" represented for them:

> The construction of situations begins on the ruins of the modern spectacle. It is easy to see the extent to which the very principle of the spectacle—nonintervention—is linked to the alienation of the old world. Conversely, the most pertinent revolutionary experiments in culture have sought to break the spectator's psychological identification with the hero so as to *draw him into activity*...The situation is thus *made to be lived* by its constructors. The role played by a passive or merely bit-part playing "public" must constantly diminish, while that played by those who cannot be called actors, but rather, in a new

sense of the term, "*livers*," must constantly increase. (*Report on the Construction of Situations* in Knabb 1981, 43, emphases added)

Situationists wanted to "reduce the empty moments of life as much as possible" (Debord in Knabb 1981, 23–4) by inventing games that could disrupt and bring joy back into the banality of modern life. They saw what the masses were doing, and what European city plans and the spectacle were doing to the masses, and wanted to create alternative adventures that would draw individuals out of their behavioral routine. They wanted to create moments that would allow and encourage individuals to experience, see, and live in the spaces they were trained to simply move through. While not explicitly promoting wandering or a modern form of flânerie, Situationists critiqued the absence of life within Parisian city space and saw the general lack of adventure within the space that makes up Western society as a key reason for a society full of spectators.

As central as the city was as the terrain for Situationists actions, their playful anti-ideological and explicitly disruptive practices went beyond momentary life jarring surprises in the city. For example, Situationists resisted the ownership of language and ideas by playfully rearranging and resituating already existing ideas in new and unthought of ways. Referred to as détournement, this action was used to draw attention to the political significance of freeing language and images from their normalized meaning.[41]

Détournement is a sort of reuse of terms in such a way as to adapt them to particular strategic needs thus freeing them from their assumed place. The point was to transcend "the bourgeois cult of originality and the private ownership of thought" (Jappe 1999, 59) and allow language to be used by temporal and spatial travelers for whatever means they may have. More than simply using language differently, Debord wanted to challenge the entire notion of the ownership of language. He and his fellow Situationists enacted this challenge by bringing together divergent objects and creating new relationships among objects that may have no organic relationship to one another. The intent was, once again, to turn the city around and uncover ways of replacing the poverty of everyday life with empowering and unifying opportunities that could be experienced once the city was seen as an adventurous place rather than a space of consumption and passivity. Détournement is explicitly playful and disruptive, and when combined with the construction of situations and his theory of the *dérive* (literally the French word for drifting), it was Debord's way of publicly creating life as a work of art as a response to (yet also impervious to) the conditions of modern life. He and the Situationists brought to light, challenged, and disrupted the predictability of individuals whose desires and needs were fed to them like any other commodity. The Situationists saw modern capitalist society as a fixed moment conserved by an organization of spectacles that obscured the real, which was and is never constant. Debord's focus was less on changing the spectacular system than on uncovering it and thus waking the masses to their condition.

Debord's Theory of Dérive is of particular relevance to the defense of wandering occurring in this chapter. A dérive is a spontaneous event whereby "one or more persons during a certain period drop their usual motives for movement and action, their relations, their work and leisure activities, and let themselves be drawn by the attractions of the terrain and the encounters they find there" (in Knabb 1981, 50). It is to be "a technique of transient passage through varied ambiances [which] entails playful-constructive behaviour and awareness of psycho-geographical effects." This awareness, Situationists believed, "completely distinguishes" this type of drifting "from the classical notions of the journey and the stroll" (in Knabb 1981, 50). The dérive was a way of moving through space in such a way that one's surrounding took on agency. A dérive was always directionless so as to ensure the drifter could be aware of the effects the space was having on their emotion, direction, and experience. Awareness of psychogeographic effects means awareness of the emotional and psychological impacts of one's environment or cityscape. While I believe the earlier example of the walking woman suggests the dérive and the aimless wander are much closer to one another than Debord and Knabb admit, it will become clear that the political moments feral citizens are intent on creating are inspired by reasons very similar to those that inform the dérive, détournement, and the construction of situations.

By initiating moments, feral citizens play their own games and follow their own rules. Like situations, the political moments created by feral citizens will tend to be informal, short lived, and small or minor. They require and create participants and rarely if ever attract (or wish to attract) the spectacle's interest. Feral citizens would largely accept the claim that "there can be no freedom apart from activity, and within the spectacle all activity is banned" (Debord in Jappe 1999, 24), so feral citizens go elsewhere to be active.

The opportunity to bypass and/or subvert the spectacle is one of the many areas where being a feral citizen is a promise and a limit. More often than not, as I have mentioned, given the extent of the spectacle, particular desires or needs cannot be satiated without the support of the state, economy, and/or media, all of which are essential parts of the spectacle and make up the typical dynamic that embodies it. At times, individuals will need to perform in ways that do not conform to the methods and requirements of feral citizenship; one more reason why feral citizenship is not intended to replace or encompass the diversity that already makes up the democratic public sphere.

The examples I have given of literal walkers and defenders and practitioners of the wander offer ways of acting within and reacting to modern banalities. In what follows I look at certain particularly threatening conditions modern political wanderers—literal and metaphoric—face within modern spectator society.

Social Movements and the Pilgrimage

As limited as any dictionary definition inevitably is, when it comes to the act of wandering the *OED* does offer important hints at how the act, when grafted onto political subjectivity, can be democratically useful. According to the *OED*, to wander is "to walk or move in a leisurely, casual, or aimless way. Move slowly away from a fixed point or place. Move slowly through or over (a place or area)." In this section, I argue that slow speed and aimlessness are two key attributes of wandering that distinguish citizenship activity from the end-oriented activity of critical social movements such as ecology, feminism, gay and lesbian politics, and anticapitalism, to name a few of the most active among them.[42]

The pilgrimage is the closest walking cousin of new social movements, and pilgrimages, as important as they have been and continue to be for many, are different from wandering. During a pilgrimage, walking is usually a means to reaching or achieving a predetermined end point. As Rebecca Solnit (2000, 50) explains, for the pilgrim, "to travel without arriving would be as incomplete as to arrive without having traveled."[43] For a wanderer, travel is the *means and the end*. A pilgrimage is "one of the fundamental structures a journey can take—the quest in search of something, if only one's own transformation, the journey toward a goal—and for pilgrims, walking is work" (Solnit 2000, 45); wandering is another and distinctly different structure.

The pilgrimage and its more forceful sister, the march,[44] are purposeful acts that use peripatetic methods often used to make particular claims or draw attention to particular injustices. To be sure, pilgrimages of the past and present have important contributing roles to play in the expansion of politics. They bring people together "out in the streets" or in the countryside and they create a space of tension and attention; they generate moments or situations that cannot be easily ignored and they stimulate discussion among onlookers and themselves. Some marches, like the purposeful and pleasurable "Reclaim the Streets" gatherings, actually become celebrations of the joy of wandering, and many others likewise end up bringing people together to realize the extent of their power when they gather together to act in concert. To ignore or downplay the relevance of these walks would be a mistake, but to fail to distinguish between the pleasurable wandering of feral citizens and the necessary and end-oriented focus of most pilgrimages and marches would be equally erroneous.[45] Furthermore, wandering is rarely part of the spectacle, while marches and pilgrimages typically occur within the integrated spectacle as they usually make demands to the state, rely on media coverage, and in one way or another relate to the economy.

Keeping political (wandering) and social (marching) activity separate is particularly important for the political sphere as "it is only by respecting its own borders that this realm [the political sphere or realm of action], where we are free to act and to change, can remain intact, preserving its integrity and keeping its promises" (Arendt 1968, 264). From the perspective of social needs, the political sphere is incapable of offering the results that are rightfully being demanded as a part of liberal democracy's promise of freedom and equality. Political acts and deeds are performative and exhausted in the moment itself, therefore rather than being judged by their just outcomes need to be judged for their originality, disruption, and life-affirming joy.

Likewise, acts of wandering cannot (and ought not to be required to) be defended against particular social needs, but neither should wandering attempt to encompass the necessary acts of resistance and demands for equality and the elimination of oppression that stimulate most social movements. It is less that movements are antipolitical than it is that the political aspects within such movements are obscured and need, at times, to be allowed back in. Whatever the particular movement goal may be, it will always need to return to political wandering and discussion.

The privileged absence of needs and the related opportunity to explore and venture throughout wild terrain for the pleasure of the act/wander may appear exclusionary and frivolous in times of need. Yet ever-present needs (constructed or real) rely on activities and rules that are distinctly different from political activity. Procedures and deliberations must be efficient and fair but they are not intrinsically valuable or creative. Though no less relevant than political wandering, the response to social needs must be seen as essentially different. Once again, it is not that movements are necessarily and fundamentally apolitical, it is rather that movements often fail to consider the fact that their movement relies on a political culture that is threatened by the absence of political engagement. Too heavy a focus on demands and too little attention to democratic ideals and struggles are making the agora where social and political desires can coincide and struggle, into a sphere dominated by administrative demands.

Feral citizenship is an intentionally particular and marginal way of interacting in political space. The idea, to repeat a key point, is not to destroy all those communities and homes that presently exist. The idea is to incite and politicize communities, to democratize them, and to open them up to political opinion in order to allow for and encourage what Nietzsche might call perspectival thinking within their community where everything that was once considered true would be reconsidered as a particular perspective that may be truthful but is not truth.

An intended impact of feral citizenship is a replacement of the banality of existing politics with playful, wonderful, and vibrant performances and moments that entice and excite democratic citizens at the same time as they uncover the irreducible need for more and freer democratic terrain. In the case of social movements and political organizations, the purpose of feral citizenship is to remind actors what sort of culture has allowed them to flourish, and in doing so, show them the need to protect and value political engagement that is free from predetermined or specific ends. Perspectival thinking creates the conditions that can lead to perspectival engagement with others, which in turn can uncover the importance of plurality, public engagement, and active listening.

In ideal situations, dynamic public spheres are the result of curious democratically guided citizens who *without intermediaries* communicate with each other out of the joy that comes from debate and performative interaction among passionately opinionated citizens who want a say in their community's activities and ideals. These citizens extend and occupy the political sphere in order to embrace and act out their freedom.[46] In less-than-ideal situations where active citizenship is not as prevalent and democratic public spheres are not as full of vibrant activity, particularly committed democratic citizens such as feral citizens become useful if not essential as they can help populate the political desert with oasis, situations and adventures that might be enough to entice more people to wander the democratic terrain that may be obscured but has yet to be paved over. Democratic terrain needs to be brought back to the attention of those who pass through it as they all depend on its presence and many have forgotten how creative, playful, and liberating a stroll can be.

In this chapter, I have shown how wandering can be an important component of democratic citizenship. By focusing on the relevance of individual experience and exploration, and in particular by showing the importance of traveling and the need for interesting places to travel, I have argued that both physical and theoretical wandering are valuable components of active citizenship within healthy democratic societies. Yet, there is something missing. To be democratic and to be radical wanderers who stimulate political discussion and create political space, citizens must also be engaged and disruptive. One way to do this is to be consciously feral.

Chapter 2

Why Feral

Feral animals[1] are generally described as domestic animals that have returned to a wild state. Typically, these animals are viewed as invasive species that destroy the natural or native makeup of the environment they enter. Managers debate over the best way to eradicate or control such species in order to allow the environment to recover or be restored to its "appropriate," predisturbed state. The numerous levels of cultural and racial bias, the degree of arrogant anthropocentrism, and the assumed "needy nature" that such helping professions and conservation strategies often carry, deserve a lengthy study of their own but that is not the focus of this chapter. Biases and limitations of the knowledge that emerges from such assumptions are briefly considered, but my main interest lies in defending a political methodology or approach to citizenship that relies on the term "feral" to describe the activity and subjectivity of political agency.[2]

The justification of turning to the feral metaphor—in order to celebrate the image of a hard-to-identify border character who disrupts and transgresses political taxonomy and ways of acting or knowing—goes beyond finding democratic promise in lessons learned from social reactions to the presence of feral animals. There are also many cases of wild children who have been "discovered" and brought into human society. The "discovery" of human ferals, while less physically disruptive to the environment, has always been immensely disquieting to social assumptions concerning what it means to be human.

The mere presence of a feral child instigates responses and reactions that are both fascinating and frightening. More importantly, at least when discussed in relation to political promise and citizenship, feral children, by no intent of their own, disrupt many of the taken-for-granted beliefs of all those who interact with them. Feral children are rarely viewed as those from whom society can respectfully learn. Neither, as Barbara Noske (1997) explains, are the animals that care for feral children ever looked at with the admiration and respect they deserve.

My intention here is neither to defend nor to critique any particular strategy or approach to interacting or dealing with feral animals or feral children. What I do, rather, is defend the political promise of claiming the feral as a defining feature of a disruptive approach to radical democratic citizenship. Specifically, I suggest there is democratic promise in grafting the essentially disruptive traits of feral animals, be they human or not, onto the political subjectivity of the citizen. I reclaim disruption[3] as an important political act and suggest that a conscious and purposeful feral identity can be a wonderful way of resisting any trend toward claiming citizenship for extrapolitical desires.[4]

Before I explain the political promise of the ideal of the *feral*, it is first important to clarify that when used here, feral and wild are not synonymous. In this book feral is a border or boundary term describing a creature that is neither wild nor domestic, a creature that may at times have been both but is at present neither and thus does not fit easily into any simple and comfortable category. The animal tends to be seen as *out of place* and thus disruptive. It also tends to be viewed or described as a problem that must be "dealt with" in order to maintain the health and balance of the community. In this kind of conservationist-managerial discourse, the moment an animal is defined as feral it loses relevance or individuality outside the context of the imposed classification.[5] The animal is, almost without exception, categorized as invasive, a pest, and a disturbance.

How eradication is to occur is typically up for debate but eradication itself is an unquestioned necessity. As the animals fail to fit into a determined taxonomy, and fail to act according to what experts determine others in their species naturally do, they soon find themselves charged as threats to their newly adopted environment they were never "supposed to be" a part of. Considering the cultural and environmentalist focus on wildlife management, conservation plans, and restoration ecology, it is not surprising that specialists as well as laymen would conclude that eradication or management of such animals is the only *reasonable* option.

A few examples of the language used to vilify these unfortunate creatures offer a good indication of just how frightening their presence is to experts such as conservation biologists. From the Missouri

Department of Conservation (http://www.conservation.state.mo.us/landown/wild/nuisance/hogs/) we learn,

> Feral, free-roaming hogs degrade wildlife habitat, compete directly with native wildlife for food, and can pose a threat to humans and domestic livestock through the spread of disease... Some of these hogs have escaped from captivity. However, some have been intentionally released on public lands for hunting purposes, although it is illegal to do so. These hogs pose a very real threat, and if left unchecked, their numbers can expand rapidly. You can help.

The way to help is to shoot as many as you can, as the specific section on Hunting Feral Hogs explains, "Feral hogs are not native to Missouri, and can be taken in any number at any time. Before shooting, however, be certain the hog is feral and is not escaped livestock."[6]

The management and control of feral cats is slightly less blatantly violent due to the fact that many cats are "domestic." Nevertheless, they remain defined as "problem species" and are usually considered pests or invasive. In fact, when it comes to felines, if one is so inclined one can take an online course, Trap-Neuter-Return, Managing Feral Cat Colonies. The online outline of the course claims,

> This course will provide you with an answer—trap/neuter/return, or TNR as it's popularly known—is the only method proven to be both humane and effective in controlling feral cat populations. Whether your focus is on the cats who've taken up residence in your backyard or on the ferals throughout your city, TNR can stop the cats from reproducing and eliminate much of the nuisance behavior often associated with ferals.

Or one can turn to Animal Rights Canada's site (http://www.animalrightscanada.com/feral/web) for the more compassionate approach. On this site we are told, "Feral cats deserve our compassion and protection. Cats, whether feral or domestic, deserve the right to be recognized as a unique and important species and to be treated as equal members of the animal kingdom." However, when it comes to what to do with the increasing population of feral cats, Animal Rights Canada unequivocally agrees with the Trap-Neuter-Return policy suggesting they can be a part of the animal kingdom only if they behave appropriately and do not overpopulate it, whatever that may mean. And at no point do any of these responses consider the issues of originality, purity, nature/nurture, nurture/nature, responsibility, disturbance ecology, conservation, adaptation, static nature, management, and numerous other complexities that are associated with the social creation of ferals.

These are just a few examples of the many management and eradication strategies that attempt to tackle the feral animal problem.[7] As mentioned earlier it is not my intention to debate the merits of the management approaches typically taken. Rather, without ignoring the language of objectification and the social bias that often informs the objectification, I defend the political promise of the focus of attention and disruption that the presence of ferals creates. Specifically, I suggest the term "feral" be claimed as a disruptive qualifier for radically democratic citizenship.

The main reason the idea of the feral is worthy of political consideration lies in its disruptive nature. Once attached to citizenship and introduced as part of a wandering political epistemology, the disruptive promise of being feral can be a radically democratic way of approaching the pluralist condition of modern democracy. The unselling the feral citizen brings will be democratic and temporary. The activity each feral citizen performs will open up communities and reinvigorate democratic discourse and public discussion, but as wanderers, feral citizens themselves, will not remain in any one space for an extended period of time.[8] Furthermore, the disruptive activity of a feral citizen is not intended to destroy the visited community, it is merely intended to agitate, and by doing so, create discursive moments that can orient the community toward seeing its activities in relation to broader democratic goals of freedom and equality.

Feral Children

> Victor made Itard as much as Itard, Victor. Genie made Curtiss as much as, or perhaps rather more than, Curtiss affected Genie. (Leiber 1997, 326)[9]

The feral child represent the unwilling embodiment of the age-old forbidden experiment that would allow scientists to observe what, if any, language or other human attributes develop when humans are in a "pure state of nature."[10] Feral children are frequently as objectified as their nonhuman counterparts are, but they tend first to be pitied as children in need of being saved, followed quickly by the belief that if saved they may have secrets to share. Due to the human (culture) and inhuman (nature) ambiguity they represent, feral children become objects of fascination. What they lack in *human* faculties, they are presumed to gain in extraordinary and distinct attributes;[11] the trick or goal of the observer and/or scientist is to find out how to tease the secrets out of a noncommunicative subject. Notice the move from compassion or help to self-interested objectification or (ab)use.

According to philosopher Joseph Leiber (1997, 330), "the question which burns in the minds of those concerned with wild/isolated children" is the one that will lead the child to answer the question "*exactly what it was like for you to be a wild/natural child*" (emphases original). Once it is recognized that the child is incapable of revealing this secret—as is always the case—the interest in the child tends to wane. The well-known cases of abused and isolated Genie, and Victor of Aveyron are two examples of such unfortunate and unforgivable endings. Sadly, social prejudices along with anthropocentric arrogance are not limited to the originating gaze and categorization. Throughout all stages of the research, the desires of the individual scientists, society, or the community consistently outweigh the needs of the child.[12]

Michael Newton begins his book *Savage Girls and Wild Boys: A History of Feral Children* by suggesting that the cases he is going to reflect upon are all "tales of pursuit" (2002, xviii). The discovery of feral children, he explains, has always been an opportunity, a promise for discovery and a promise for self-advancement for those fortunate enough to participate in the rehabilitation of the abandoned, placeless child. In Newton's words,

> The pursuers are various, the young French surgeon flushed with the possibilities of a great enterprise; the wearied, sardonic Scottish doctor; the village priest willing to believe; the eccentric judge; the gentleman of leisure; the errant aristocrat. Yet the object of their pursuit is constant. All of them seek the truth, one that is embodied in another human being; and, for each one, that truth is something that can only be found in the exceptional fate of *this* boy, or of *this* girl. That is the end of their quest, to fix for a moment the fleeting truth glimpsed within the life, the eyes, the soul, of the wild child. (2002, xviii)

Feral children are situated in particular times and particular spaces; so too are they rendered valuable through particular expertise and disciplines. Victor of Aveyron, for example, entered post-revolutionary French society at the same time as Rousseau's noble savage was the main topic of discussion within the French salons. In fact, the training of Victor was at least partially undertaken so he could be displayed to Madame Recamier[13] (Newton 2002, 100) and her *important* crowd who wished to observe an actual "noble savage." But unfortunately for Victor, he never became the teller of the truths scientists were searching for and his incapacity to perform the role prescribed to him by Jean-Marc-Gaspard Itard and other less vigorous researchers meant his presence became less and less helpful to their own projects. Like nonhuman ferals, Victor's uniqueness and oddity was never valued or respected in and of itself. His (or perhaps its) value only mattered in relation to those pursuing their opportunity to use Victor to get to some new truth and thus make them famous. Once his usefulness to the researchers ran out—he stopped showing improvement and making Itard (his scientist trainer and savior) proud—he was left to Madame Guérin's[14] care and no longer considered useful to science or society. As Steeves (2003, 230) points out, "it is without question that when we study feral children we inevitably learn more about ourselves than our subject."

The sad and exemplary case of Victor "The Wild Boy of Aveyron" as well as the extent to which we can learn about ourselves is made more evident in J. J. Virey and Pierre-Joseph Bonnaterre's comments on their experience of working with him. These two scientists were frustrated with what they interpreted as Victor's self-centeredness and inability to adapt to social norms. They both observed his indifference to kindness and his inability to show gratitude. Virey asked,

> Would you look after him? He does not concern himself with anyone in the world, not even who feeds him…Would you give him something to eat? He grabs it immediately without indicating the slightest gratitude. He thinks of nothing, or rather he feels nothing but his self alone, he is the invisible unity, pure egoism; he attaches himself to no one, to no creature in the world; he recognizes his guardian because he gives him [food] to eat, because he attends him, but he has no affection whatsoever for him. (Yousef 2001, 254)

Sure, the feral child who has been first abandoned, then "saved," then poked and prodded by self-interested scientists, may have a hard time with trust, and—given previous lack of human interaction and a concentrated focus on survival—may appear to be self-interested. In fact, it is hard to imagine any other response to his new situation. Learning more about ourselves, we see that the focused scientists seeing the child as a means to an end, are the most self-interested, yet unlike the feral child they cannot claim as cause past abuse or a lack of understanding of human compassion. Forgiveness for the scientists, if any is justified, lies in relation to social pressures and personal desires that inform scientists who are situated in a particular social imaginary that encourages and/or requires them to view the child as a means to an end. If the end is not achieved or achievable, the child or object of study is to blame not the scientist or scientific method.

Indeed, blaming particular scientists or researchers may not accurately portray the cultural, racial, sexual, and professional prejudice feral children and other oddities inevitably face. The gaze of the researcher is as much a result of society's desires as her/his own properly distanced observation and experimentation. The prejudice of disciplines suggests that to uncover and discover truth, scientists and researchers are to be distanced and unemotional.

While love and care are undoubtedly most important to the child, they do not entice those who "visited promptly" to have an opportunity to examine "a child cut off from all of society and all intellectual communication, a child to whom no one had ever spoken and who would be scrutinized down to the slightest movements" (Yousef 2001, 247). Itard, who "left us a story not just of the education of a young savage, but his own inability to be close or intimate" (Newton 2002, 127), astutely noted that Victor was far more friendly with Madame Guérin, the housekeeper who cared for him, than he was with himself (Leiber 1997, 328). Itard was aware enough to acknowledge that Victor's feelings for Madame Guérin were due to the care and lack of instruction that she gave Victor. Guérin was the only companion Victor had who did not see him as a means to an end, she saw him as an individual and as such she gave him love and care, something he may or may not have had before being "discovered."

SELF-REFLEXIVITY

Whether mythical or real, the stories of nearly human animals and nearly animal humans "challenge the boundaries of our communities in many ways, forcing us to ask questions of our collective identity and the ways in which we experience ourselves in the world" (Steeves 2003, 231). "What is clear," Steeves (2003, 244) goes on to explain, "is that the comfortable fiction of a human/animal dichotomy and the notion of a strict definition for 'human' and 'animal' are threatened by feral children."[15] Feral children, not by any fault of their own, embody a challenge to those who believe they have "got it right"[16] when it comes to defining what counts as human. Humans may well be distinct from other animals but so too are they similar; it is the similarities that most disturb anthropocentric scientists, and it is the apparent similarities that have driven them to struggle to come up with a definitive answer to what makes humans distinct from other animals.[17]

For those who agree with the argument that disruption and critical interrogation are essential to a healthy democracy, the disruption and disturbance latent within the metaphor of "feral" can be seen as a useful political qualifier for a democratically informed citizenship. To paraphrase Steeves, I hope the very existence of feral citizens will be a persistent threat to our understanding of what it means to be a citizen. As a method for performing political agency, feral citizenship becomes, resituates, and embodies much of the forbidden experiment. However, the feral citizen chooses the life of a feral and the forbidden experiment is more ideological than physical. The feral citizen is politically homeless and does not fit in any representable category, but s/he does not see this as a problem. On the contrary, homelessness and nonrepresentability are compulsory conditions for the radically democratic activities of feral citizens.

In fact, is not democracy in some way humanity's collective forbidden experiment? Are those of us living in Western democratic societies not at least rhetorically living in a world that no longer has singular guiding truths and authoritative rules? The tribunals of history are no longer consulted in order to guide and legitimize our choices and discussions. There is no longer a common understanding of the good life to inform and teach how to act appropriately and there is no longer a truth-teller to put us right. Our collective forbidden experiment concerns what humans can become once freed from truths, disciplines, and metanarratives. This is, again, why being a consciously feral citizen is so potentially disruptive and exciting. The explicitly political ideal is to use the disruptive potential of an indefinable, partially wild and partially domestic, homeless, and uncategorizable citizen to ensure the disruption is maintained and the experiment continued.

Democracy, if taken seriously and accepted as an ideal, requires critical engagement with one's surrounding. It requires disrupting and challenging the norms that guide particular communities, and it requires using the undisciplined political gaze to uncover the conservative and limited perspective of those who have become comfortable in their role and position in society. But, just as the wanderer needs places to visit, so too does the feral citizen need wild and domestic places to disrupt and challenge.

Ferals and the Pure State of Nature

The idea of a "pure state of nature" is a curious one.[18] We know that in Victor's case the pure state of nature was considered a solitary state due to the prominence of Rousseau and his belief in a presocial noble savage,[19] and we know that if the association of feral and isolated children with each other is at all justifiable it is because feral and isolated children both lacked human contact. Yet as Steeves (2003, 245) correctly suggests, one thing we find from examining feral children is that "humanity is in some respect the result of specific treatment within one's community. To have human experiences, one must be attended to as human. To develop human intentionality one must be treated as if he or she already possessed such intentional structures. Being human is being treated by humans as human." The same can be said for citizens who are presently treated as "tax payers," "stake holders," "constituents," "property owners." As citizens are addressed as these identities, these identities are being internalized and referred to for legitimacy and the right to speak within political discourse. But more sinisterly, as citizens are no longer being addressed as active political agents of change, they are losing their capacity to act as political beings.

If there is a politically salient conclusion to draw, it is that human potential and human growth require active engagement with others. The search for knowledge about human nature from a child in a so-called pure state of nature is futile as the distinctness of a child lies not in the presence but in the lack of autonomous purity. The absence of human interaction does not give us a state of nature; it gives us a state of isolation, sorrow, and need.[20] Why else would such conditions be part of a forbidden experiment? On a societal level, it should be noted, isolation, sorrow, and persistent real and constructed need are key components as well as outcomes of neoliberalism. It is thus not uncommon to find "dumb, slovenly, incurious and unresponding boy[s]" (Newton 2002, 100) in modern times where individualized conditions of the forbidden experiment have obscured the promise of the freedom granted.

A human "pure state of nature" is a construct that allows the ghost of presumed human homogeneity to resurface, and resurface it has. The attainment of something like Rawls's original position remains a desire for numerous liberal theorists and the disembodied, rationalist, and unsullied voice of the public participant remains a regulatory requirement for citizen activity within much of deliberative democracy's procedural focus. For many democratic theorists, the attainment of something like a pure state of politics is a lot like the pure state of nature that enticed the researchers of feral children.

Like researchers of feral children, theorists searching for such a pure politics or a pure political agent fail to realize the degree to which plurality and difference are constitutive of the democratic condition. They also, like those struggling to define what humanity is, fail to acknowledge that any definition of humanity or the ideal citizen is particular, often sexist, and almost always racist. Steeves (2003) and Newton (2002), in their discussion of cases of feral children, outline numerous racist and sexist biases that reappear when the category "human" is disturbed by the presence of a feral child. Rather than go over the many examples it is fair simply to accept the point that whenever humanity is described it fails to include *all* humans. The presence of feral children reintroduces the failure of any definition and forces a reconsideration of how we define humanity. Any definition rests "on a multitude of unarticulated assumptions" (Steeves 2003, 233), most of which are entirely unjustifiable and thus in need of being challenged and constantly revisited.

Wandering feral citizens are embodied political agents who *create* political space and challenge fixed knowledge, isolation, and constructed needs through their life affirming methodology. Ferals, as we have learned, cannot help but disrupt comfortable and closed definitions, so when ferals become political creatures they are almost by default radically democratic. When feral animals enter environments, or when feral children are discovered, they become a threat to the tidiness and permanence that has helped define the entered environment. They do not destroy the community they enter but they certainly disrupt it and bring to light unexamined assumptions and beliefs.

Feral citizenship includes a passionate desire to extend politics and democratic culture back into those communities that exist only as a result of democratic culture in the first place. The intention is that by repoliticizing spaces, and then rambling on, communities will themselves open up to the political sphere and begin to recognize the significance of *also* contributing to the tension-filled sphere of the democratic public sphere.[21]

CHAPTER 3

Why Citizenship

The way to render wandering and feral characteristics politically significant is to graft them onto the citizen, the subject who renders the political sphere a living reality. This grafting is important for two key reasons. First, the conscious activities of wandering ferals could very easily devolve into nihilistic destruction without recognition of the primacy of the political and the importance of critical interrogation and disruption as a specifically political act. Second, as Susan Bickford (1996, 186) has pointed out, "no one can be actively engaged in the tension of citizenship all the time, or even most of the time, and politics is not the whole of human existence." Thus, feral subjectivity and methodology as part of citizenship is defended in this chapter as a conscious and or temporary choice utilized by those committed to expanding the public sphere and revitalizing the democratic tradition. Feral citizenship is not a type of political agency for nomadic peoples; it is an approach to political agency intent on encouraging and celebrating disruptive, nomadic agents of politics.

The fashionable use of adjectives—green, cosmopolitan, queer, multicultural, feminist—to help ground or claim citizenship for particular purposes is understandable given the prevalence of the rhetorical commitment to democracy, but without an approach to citizenship that remains explicitly political as far as not being enticed by achievable ends, such purposeful and fixed use of citizenship threatens to gloss over the necessarily democratic commitment of all citizens(hips).[1] Feral citizenship, while it does not solve the problem of citizenship diversity, does attempt to bring it to light and by doing so render it a topic of political discussion.

So why propose another theory of democratic citizenship? Why further muddy up the already murky waters of citizenship studies? First of all, I write this chapter because no one has got it right (and no one ever will get it right). Every approach to citizenship excludes some voices, some ideas, and some individuals. The continuous extension of politics into previously unthought domains is an indication of how many relations have yet to be politicized and how much previous theories of democratic citizenship have missed.

In the introduction to this book, I mentioned that the best and most "democratic" definition of democracy, the one that best reflects an ethicopolitical commitment to democracy, is offered by Anthony Arblaster.[2] According to him, democracy is

> not only a contestable concept, but also a "critical" concept that is a norm or ideal by which reality is tested and found wanting. There will always be some further extension or growth of democracy to be undertaken. That is not to say that a perfect democracy is in the end attainable, any more than is perfect freedom or perfect justice. It is rather that the idea and ideal is always likely to function as a corrective to complacency rather than a prop to it. (1987, 6)

In approaching democracy in this radically participatory and critical manner, a healthy democracy becomes a playground for critical citizens[3] who, through initiating, performing, telling, and listening to stories, nurture the plurality of public spaces. As a citizen in this sort of active democracy, one has specific political responsibilities to a diverse group of others, not the least of which is ensuring the continuation of the condition of plurality, and by consequence the primacy of the political. Plurality is a necessary though not sufficient condition for democracy and as plurality is encouraged by unpredictable and undisciplined citizen activity that likewise depends on and emerges from plurality, adding to and disrupting current theories of citizenship seems like a worthwhile endeavor for those interested in democracy.

In the previous two chapters on wandering and ferals, I argued that feral citizens are aimless wanderers of political terrain. Their intent is to visit and disrupt communities and other individuals they meet along the terrain. The intention is not to entice those met to abandon their homes and join the feral citizen but rather it is to remind those with homes that they live upon democratic terrain and without such terrain their home, their community, and their liberty is lost. Thus, while feral citizens do not persuade others to follow them, they do want to convince their interlocutors to go for strolls and experience the terrain, for it is along the terrain where those who occupy homes may become aware of the diversity of homeowners and renters along their street and throughout their community. While a particular concern may have helped them build their home, the realization of a diverse community may help them appreciate that the value of their home is depreciated if the community faces a democracy bubble and houses are foreclosed or renters are forced out. While feral citizens may squat in the foreclosed homes, they also remind all homeowners of the importance of community and democracy.

It is as citizens that we can and must risk the dangers associated with interacting with, learning from, and persuading others through means other than force. For the citizen, "not life but the world is at stake" (Arendt 1968, 156), and that world is guided by the distinctly human potential to step beyond mere survival, to be guided *by* freedom and to *become* free through the act. A politics-first position such as the one defended here requires democratic ideals of freedom and equality inform the actions of the citizen who "should respect pluralism and individual liberty" and thus recognize that "every attempt to reintroduce a moral community, to go back to a *universitas*, is to be resisted" (Mouffe 1993, 56). For feral citizens, the act is less a risk than it is a playful (and serious) wander along the political terrain that s/he dwells in, creates, and nurtures. The risk comes to those who are less willing to fully embrace the politics-first position but are nevertheless committed to democracy enough to actually listen to the arguments of those who are convinced.

Democratic terrain is both created by, and thrives on a tension that evolves out of debate and discourse nurtured by "a specific language of civic intercourse, the *res publica*" (Mouffe 1993, 63), internalized by active political agents. By assuming a set of rules that requires acknowledging and embracing the plurality of opinions that populate political terrain, citizens embody an ethic that guides democratic involvement. At a minimum, this ethic includes nonessentialism, recognition of fallibility, and respect for the positions of others. "Only this common acceptance," Mouffe (1993, 63) asserts, "can lead to a shared political identity among people who are otherwise part of numerous unconnected activities." Thus, when acting as a citizen one must embody a particular way of interacting that limits as much as it liberates the sort of discourse that is deemed acceptable. Feral citizens remind all other citizens that citizenship is about politics, opinion, debate, action, and discourse. As just mentioned, when directly engaging with others, feral citizens are not interested in creating more feral citizens. They are, however, interested in creating more active citizenship and encouraging others to recognize the priority of the political. By embodying and being committed solely to radically antiauthoritarian democratic ethics, feral citizens are like roaming informal and marginal political spheres, creating political space and political subjectivity wherever they go.

Public realm theorists from Arendt to Castoriadis to Mouffe to Habermas all recognize the need for distinct spaces or forms of interaction that are not subject to temporal and spatial needs. They certainly disagree on the purpose for these spaces but what each is certain about is the irreducible need for *coercionless* interaction among equally valued citizens. Feral citizenship is a particular type of political agency that embodies such beliefs; as an active political agent wandering along, participating in, and creating democratic terrain, the feral citizen's primary role lies in assisting in ensuring such spaces continue to exist and flourish.

Before I conclude this brief introductory defense of citizenship as the appropriate common identity to attach the characteristics of wandering and feral to, it is important to reiterate the disclaimer that citizenship and politics are only ever one part of one's life, not all of it. Wandering feral citizens are not always and only political agents, and even as political agents they are not always wandering and feral. Actors may always have the potential to become wandering feral citizens, but this is not the same as always being political. One need not be constantly political and permanently engaged to be a feral citizen, but by bringing together the aimlessness of the wander, the disruptive nature of a feral identity, and the politics-first characteristics of the citizen, we have all the ingredients needed for active political agency and democratic rambling.

The Many Functions of the Wandering Feral Citizen

It is hard to decipher whether, for Arendt, the actor, spectator, and storyteller—all necessary for the construction of political moments—are fixed or fluid identities. Wandering feral citizenship leaves no guesswork; the citizen can be all these identities, and can participate in the realm of appearance as a perpetual amateur, taking on at times each of these political identities.[4] A wanderer who is out in public—not only off the treadmill but also off the beaten trail—can be a spectator, a storyteller, and an actor. By disturbing the surroundings, s/he creates, or stimulates the creation of political moments where individuals appear in front of an audience, where opinionated actors perform for each other and create the stage for the unexpected to appear. As each political character(istic) is viewed as an activity rather than an identity, all situations or encounters offer the opportunity for each of those present to be actors, storytellers, and spectators all in the same place and in the same political moment.

Referring to the fragility and spontaneity of Arendtian inspired politics, Maurizio Passerin D'Entrèves (1994, 77) argues that spaces of appearance must be continuously recreated through active participation, "Its existence is secured whenever actors gather together for the purpose of discussing and deliberating about matters of public concern, and it disappears the moment these activities cease." This always possible, yet never fixed space is only actualized when individuals gather together to participate and act in spaces free from rule or authority; acts that are much easier to stimulate by those who refuse categorization and celebrate disruption.

Arendt (1958, 197) argues, "The calamities of action all arise from the human condition of plurality, which is the condition *sine qua non* for the space of appearance which is the public realm. Hence the attempt to do away with plurality is always tantamount to the abolition of the public realm itself." What feral citizenship does is acknowledge that "to be free mean(s) to be free from the inequality present in rulership and to move in a sphere where neither rule nor being ruled exist(s)" (Arendt 1958, 31). This belief is embodied by wandering the political sphere and creating micropolitical moments that constitute temporary political theaters that in turn create the spaces of appearance Arendt rightly considers necessary for politics.

Included in the rationale behind Arendt's (1968, 241) defense of plurality is the opportunity for visiting others. Such visiting, she argues, not only helps the individual in "being and thinking in my own identity where actually I am not," but it also allows the visited to find the truth of their doxa that in turn allows the visitor the opportunity to understand the other's position within the common world. The "where I am not" is of particular significance as it acknowledges the need for others and other places while offering an implicit defense against the arrogance of believing that visiting can allow one to adequately represent others. This is not to say Arendt does not wish for representative thinking in much the same way as she wishes for truthfulness, it is rather to suggest that to represent another or to speak the truth is different from the obligation one has to try to be representative and to speak truthfully. The latter is based within and accepts the irreducibility of a pluralist foundation while the former threatens to silence the plurality and replace it with a new universal.

The world opens up differently to every person according to where they are positioned in it and while visiting may help one understand other places it does not allow the visitor to see as or for the visited. This realization reiterates two essential attributes of feral citizenship: First, feral citizens are only ever one kind of citizen among many; and second, feral citizens need places to visit in order to learn from and disrupt so their intent is not to encourage followers but rather to engage others and draw attention to the common terrain shared by all those committed to democracy and the ideals of freedom, equality and social justice.

Feral citizens learn and gain as much from those who they visit as the visited gain from them. Even so, for feral citizens the attainment of "representative thinking" is only one of many possible outcomes of the visiting s/he does. The effect and stimulus of the moments constructed "when citizens actually confront one another in a public space," in order to "examine an issue from a number of different perspectives, modify their views, and enlarge their standpoint to incorporate that of others" (D'Entrèves in Mouffe 1992, 152) has innumerable possibilities for *all* present and for all subject positions. Significantly, the moments must be experienced and cannot be represented, which means for feral citizens the moment not the outcome is the point of interest.

Each participant and/or subject position is (re)created through and during the event. Participation is not an option: it is a requirement if not inevitability. All subjects in political moments created by feral citizens are rendered participants as each becomes imbued with an essential role to play in the relations between all the necessary political characters (actor, spectator, and storyteller).

We know from the proliferation of social movements and the scope of the integrated spectacle that often in modern political relations the spectator and storyteller are preconstituted and assumed. The spectator tends, by default, to become the administrative body of the state as the institutions of the state have the authority to react to the demands social movements are typically putting forward. Likewise, the storyteller becomes the heavily translating and immensely powerful corporate media as it in turn has control of the primary storytelling apparatus.

The storyteller and spectator that Arendt speaks of were not *in* the performance, they were however *of* the performance. The same cannot be said for the modern-day state and the media. Both are obvious and central components of a larger performance that they themselves largely control; each has its own director role to play and at best reluctantly shows interest in performances that are not a part of their own theatrical spectacle. The proliferation of social movements and democratic eruptions around the world suggests the plurality required for creative politics does still exist, but it seems clear that the creation is not going to occur if the spectators and storytellers are the state and corporate media.

Feral citizens perform and help create political theater but their performances will always be amateur, inclusive, and spontaneous. This persistent informality will give them the ability to spontaneously and regularly create actors, spectators, and storytellers each of whom have an essential role in the transposing of human life into art. Arendt (1958, 167) correctly saw the theater as "the political art par excellence" as unlike other art forms that tended to be means to an end the performances on the stage were exhausted in the moment as the "sole subject [of a theatrical performance] is man in his relationship to others" (1958, 167). In unprompted theater, the characters and relations between performers and audience are not predetermined or institutionalized. Each emerges as a result of the moment, and in ideal political moments each participant alternately performs, listens, and tells stories as the moment temporarily frees them from the predetermined and fixed limits of their responsibilities outside the moment.

Political moments become gifts that allow and encourage one to be temporarily freed from what they are or perceive themselves to be. The moments allow the participants to act in relationships with others who have gathered together not for some kind of desired outcome but for the pleasure of the moment itself. Freed from the need to be responsible, efficient, and reasonable those present can begin to dream, imagine, and explore their capacity to be free, to act, and to unleash the unimaginable. Each political moment offers the promise of the unimaginable.

As each political moment is created by and creates amateurs in any one of the stated roles, acts of feral citizenship have the possibility of resituating and reclaiming politics outside the confines of the state and media in much the same way as the occupy movements have. In fact according to Paul Mason (2012, 84) in his book *Why It's Kicking off Everywhere: The New Global Revolutions*, one of the most obvious similarities between the diverse movements taking place around the world in the past few years is what he refers to as an almost mystical determination "to create areas of self-control" in order to "occupy a symbolic space and create within it an experimental, shared community." If Mason is right, "these attempts at creating instant 'liberated spaces' have become the single most important themes in the global revolt" (2012, 84), and the best way to understand these moments is to look at the nurturing and realization of autonomy and freedom by participants who see themselves as performing democratic moments, not primarily as means to some other end but as ends and valuable experiences in and of themselves.

Professional actors and perpetual spectators are what the typical dynamic (state/economy/media) seeks. The dynamics of spontaneous, informal, and wild citizens are undomesticated, unstructured, and unplanned making them and the participants unpredictable and uncontrollable. The theater and performance is much like street theater that appears and travels wherever it needs to or wants to; its amateurism and spontaneity signals that a potential for the creation of political moments is always present.

For representatives of social movements (what I was previously referring to as homeowners along the democratic terrain) to participate in the informal and amateur theater feral citizens initiate, they will need to accept that not everyone can always be an actor performing her/his own play, at times each participant will need to be(come) a different kind of participant. In addition, there is neither any assurance that the story being told will be interpreted as the actors hoped,[5] nor is there any reason to assume that all the spectators will receive the same message. Thus movement advocates, as potential allies, will need to realize that within the political performance there are no guarantees and to participate they may also need to spend as much time listening and spectating as they will persuading and acting.

As the interpretation of the moment, the impact of the moment, and the acts of those participating in the moment can never be controlled or predetermined, Arendt argues those who risk the political must be forgiven for the consequences of their actions. I add, here, that the political sphere must be expanded so the forgiven can continue to stimulate the forgivable. The political moments created by wandering feral citizens cannot offer those fighting for equal recognition and distribution of rights and goods an adequate platform for their needs.[6] The venue for their concerns is different and should not be thought of as being encompassed by the political moments initiated by feral citizens.

As a political methodology for amateur political agents, feral citizenship moves beyond Arendt's more purist view of politics without denying the importance of keeping political interaction distinct from other modes of engagement. It is in relation to *being political* that the greatest promise of feral citizenship lies, and it is in relation to being political that wandering and a consciously feral identity is so essential to a "politics-first" approach to citizenship. Wandering feral citizens create political moments and consider themselves capable of being *multiply political* (spectator, storyteller, and actor). Doing so offers a way of keeping political subjects with numerous subject positions multiply oriented and always interested. By being capable of becoming a spectator, storyteller, and actor feral citizens entice, annoy, challenge, assist, and ask for stories.

In the last three chapters I have shown that as a committed homeless democrat, the wandering feral citizen has a unique opportunity to participate and help encourage others to participate. The goal of feral citizenship is to disturb and entice others to create moments or situations that require active engagement. The approach of the feral citizen is primarily critical and disruptive as s/he is grounded in a belief that democracy is necessary because pluralist society does not enact a perception of the common good. In the next two chapters I discuss how feral citizenship as a methodology, and feral citizens as guides, can help revitalize the public sphere and liven up the debates that are currently defining and housing democratic discourse. Primarily, I look at how aimless wandering through democratic terrain and conscious disruption of the terrain and those who populate it offers a radically democratic way of embracing the liberatory conditions of present democratic culture.

CHAPTER 4

FERAL CITIZENSHIP AS METHOD AND FERAL CITIZEN AS GUIDE

One of the most apparent political struggles in pluralist democracies has become the creation of common "we" spaces among diverse and tension-filled groups filled with individuals with multiple identities and affiliations. Over the past few decades, feminist and postcolonial theorists have taken on this struggle and made the issue of difference and antagonism within "we" spaces central to the development of their respective, often cross-fertilizing theories.[1] Difference is now recognized by many not as a problem to deal with or solve but rather as a key attribute of the "we" space itself. The manner in which this newfound foundation ought to be addressed, dealt with, or celebrated by those wishing to weave together individuality and collectivity is the focus of this chapter.

Not surprisingly, the primary tensions between difference and commonality end up being played out on the age-old theory-praxis battleground. This is a battleground that never produces a definitive winner. What it does do for those willing to pay attention, is lead to a better understanding of the promise and limit of the particular "we" space the participants happen to be a part of during the battle. In fact, many feminists and postcolonial theorists now accept that the democratic terrain where such battles take place is as essential within their own "we" spaces (theory) as it is for the "we" spaces to act and appear in broader public arenas, where demands can be made to those with the capacity to respond (praxis).

In her book *The Human Condition*, Hannah Arendt (1958, 161) explains that "the moment we want to say who somebody is, our very vocabulary leads us astray into saying what he is; we get entangled in a description of qualities he necessarily shares with others like him...with the result that his specific uniqueness escapes us." Simply stated, when we represent we almost always represent a stereotype. If this is true, a democratic society must retain and perpetually create moments for nonrepresentable individuals to perform as who they are rather than what they are perceived to be.

Learning to Listen

Active listening has always been an implied (central) and essential aspect of discursive theories of democracy, but rarely has it been taken seriously or considered central to citizenship. Not surprisingly, the focus (and interest) of most democratic theorists has been on acts of speech and persuasion whereby the listener is assumed more than theorized, and the politics of address[2] is rarely considered. Ideals of universality or the need to persuade others of the value of a new encompassing or improved theory have consistently made active listening a responsibility of the imaginative other: the masses, the people, the victims, the marginalized, the nonhuman, the minority, the unenlightened, women, people of the Third World. This relationship has rightly been challenged and the question now becomes how can this challenge be heard by those who want to listen and learn?

According to Hannah Arendt, the moment one is acknowledged by others as a desiring subject is the moment one becomes the aforementioned unrepresentable who as opposed to a representable what. It is also the moment the individual is no longer an appropriate or expected representative of others, and thus the moment at which one can become an active, acting, performing, and indeed, irresponsible citizen. The purpose of a democratic society should be to create the conditions to allow all humans to initiate, experience, and enjoy such moments. For Arendt this is the unique promise of each and every human as we are "all the same, that is, human, in such a way that nobody is ever the same as anyone else who ever lived, lives, or will live" (Arendt 1958, 10); this is why Arendt speaks about the human condition rather than human nature and it is why the realization of distinction is found and performed within the public sphere among others who are equally capable of their own enunciative moments where what they are is secondary to who they desire to be.

Frantz Fanon sheds light on the political significance of a desiring and thus unrepresentable and unique subject when he explains that as soon as I desire I am asking to be considered. I am not merely here-and-now, sealed into thingness [representable]. I am for somewhere else and for something else. I demand that notice be taken of my negating activity insofar as I pursue something other than life; insofar as I do battle for the creation of a human world—that is a world of reciprocal recognitions. (Bhabha 1994, 8)[3]

Fanon's take on the relevance of desire, alongside his related critique of representation, lies at the heart of postcolonial criticism, which represents a direct challenge to the liberal notion of a common humanity based on multiculturalism, representative institutions, and inclusive ideologies. In its place, it offers a far more complex, uncertain, and imaginative way of viewing humanity and individual humans based on hybridity, difference, and constant intersubjective tension.

The desire Fanon speaks of emerges from those unwilling to accept the *fetishism* of identities—a translation of self-as-subject into someone else's object. Politically, it is a demand for participation as individuals, but not any kind of participation in what Arendt calls the common world. It is a demand for nonconditional or genuine participation not subject to someone else's categories or institutions of inclusion. In other words, it is a demand to participate as a who not a what. If recognized as a who, each individual is seen as incomplete, with multiple subject positions and multiple ever evolving capacities that inevitably overflow from any container that intends to give "them" a platform to speak and be heard from. Individuals who take or create the opportunity to speak and be heard should never feel obliged to represent others, as their very presence is representative of otherness.

Both the representation of others and the embodiment of otherness are important but the latter is much harder to justify in a world understandably oriented toward solutions and achievable end points. Nevertheless, celebrants of political terrain and celebrants of democratic culture need to ensure the need to represent others does not permanently replace the desire to be and perform as an other both as part of a particular we space and as part of collective humanity.

Gloria Anzaldúa follows in Fanon's tradition of demanding "respect as other" by defending what she calls border thinking, which represents a unique and personally empowering way of moving beyond a counterstance that "locks one into a duel of oppressor and oppressed" (1990, 387). Anzaldúa's defense of otherness and border thinking evolves from her physical presence on the Mexican–US border, in addition to her own personal mestiza consciousness that she believes allows her the unique opportunity to enter the world as a self-autonomous subject. For her, border thinking and/or autonomous subaltern agency involves "making face" whereby one's embeddedness is used as a way of developing unique and valuable agency irreducible to the presupposed agency typically given by the ones the face is being made to. Making face is not about mimicking the givens of place and situation; it is about marginalized individuals taking the opportunity and having the capacity to act in ways that both rely upon and resist the givens of their unchosen yet nevertheless evident place and identity.

These faces, Anzaldúa explains, are different from the masks "others have imposed on us," for such masks keep us fragmented: "After years of wearing masks we may become just a series of roles, the constellated self limping along with its broken limbs." As Homi Bhabha (1994, 175) states, "the time for 'assimilating' minorities to holistic and organic notions of cultural value has dramatically passed." The time now is for the subaltern to speak and most importantly, be heard. Bickford explains that "breaking through these masks is not, for Anzaldúa, a matter or revealing one's true inner nature and essential self; rather we 'remake anew both inner and outer faces'" (Bickford 1996, 124) by denying the fixity of all identity. The point of making face is not to find a truer self than the one imposed; it is to continue performing and developing subject positions not imposed by others, but also not entirely disembodied and free from the past present or future.

Anzaldúa announces to dominant culture,

> don't give me your tenets and your laws. Don't give me your lukewarm gods. What I want is an accounting with all three cultures—white, Mexican, Indian. I want the freedom to carve and chisel my own face, to staunch the bleeding with ashes, to fashion my own gods out of my entrails. And if going home is denied me then I will have to stand and claim my space, making a new culture—una cultura mestiza—with my own lumber, my own bricks and mortar and my own feminist architecture. (1987, 21–22)

Like Fanon, whose earlier demand in relation to desire was not to receive anything from his audience but to be recognized and respected as other, Anzaldúa does not ask for anything from those in positions of authority, on the contrary, she advises them on how to be respectful.[4] In effect she is saying if (or as) dominant culture has little value to her and those like her, allow her and those like her to be other and develop in their own way;[5] try to gaze upon them with nondefining and nonjudgemental glasses and see beyond the stereotypes that fix those like her into objects of imperialist creation. Anzaldúa explains her situation, shows and performs her agency, and declares here are the options for those who are part of dominant culture. Her agency claiming words reverse the power dynamic, she pinpoints and redefines the powerful and authoritative center and makes *it* an object of *her* gaze.

Fanon's phrase I am "for somewhere else and for something else" encapsulates the above points and implicitly defends plurality by challenging the power of core-periphery translation by informing his audience that not everything is translatable into one language and not all interests are related to the dominant one. I am not like you. I am *other* means I do not, nor do I wish to, fit your categorization. If you[6] are going to listen and engage with me in a nonoppressive manner, it requires that you start the conversation by accepting the irreducibility of certain differences at the same time as you recognize the power you have (and historically always have had) as the initiator of the conversation.

For those genuinely interested in hearing what the desiring subject has to say, they must attempt to listen without translating; they must take on the role of a spectator or background; and they must strive to achieve this difficult act all this while accepting the inevitable failure of the attempt to actually hear the words of someone who is "for somewhere else and for something else." Through the attempt both the speaker and the listener give themselves the opportunity to continue to distinguish themselves and perform the role of unrepresentable whos or perhaps more appropriately unrepresentable citizens who—through experience and noninstrumental interaction—cannot help but find the institutions of actually existing *representative* democracy wanting.

Iris Young (1997b, 354) describes the sort of active and attentive listening required as an activity that allows for the possibility that "communication is sometimes a creative process in which the other person offers a new expression, and I understand it not because I am looking for how it fits with given paradigms, but because I am open and suspend my assumptions in order to listen." The implied asymmetrical reciprocity, where one (typically those occupying the liberal core) attempts to listen without translating or assuming that by hearing the other one can put oneself in the place of the other, is a direct challenge to the ideal of representative inclusivity or the creation of ideal speech situations. It is also a burden clearly placed on the shoulders of those in positions or power as their representative legitimacy in a democratic society relies on the appearance of inclusivity and responsible governance. Susan Bickford (1996, 24) adds to Young's description of listening's creative promise by explaining that actually listening "involves an active willingness to construct certain relations of attention, to form 'auditory Gestalts' in which neither of us, as parts of the whole structure, has meaning without the other," however, she adds that such listening does not involve "abnegating oneself" as "we cannot hear but as ourselves, against the background of who we are." The acceptance of each listener's particularity means we (and by default any core) inevitably fail to understand or hear the full story of others; furthermore, as suggested earlier, full stories are only ever temporary and always relational. This is not a problem to deal with but an inevitability to acknowledge and learn from as while one need not "abnegate" oneself, one does need to be open to learning as "the riskiness of listening comes partly from the possibility that what we hear will require change from us" (Bickford 1996, 149). When the "us" in question dwells in the core that change can be very hard to accept or understand as it could go to the heart of the beliefs that legitimize the core in the first place. This is why messages from the periphery are typically translated into something that can be dealt with, understood, and accepted.

Bickford (1996, 144) is right when she argues that "in taking listening seriously we need not elevate listening over speaking as the primary political or social activity, but rather understand the interdependency, the dynamic between them, and the necessity for engagement in both modes." I think, however, the activity of listening is much more difficult and disruptive even than she seems to infer. Active listening requires those who are in dominant positions to acknowledge and take seriously precisely what they tend to assume away. The dominant "we" of which they consider themselves a part of must, at times, become an outside "them" in a constantly changing dynamic of difference and movement consisting of innumerable us-them relations.[7] Foundations, whether paradoxical or not, must all become open to challenge and must no longer be the comfortable refuge to return to and speak authoritatively from. Indeed, the entire notion of center and periphery, marginalization, and dominant culture becomes subject to ongoing critical interrogation once active listening is given its rightful place within democratic society.

Arendt, Fanon, and Anzaldúa all implicitly defend the need for democratic culture over democratic structures. They each draw attention to the incomplete nature of individuals, relationships, and communities and by doing so point to the inevitable failure of the representative institutions intending to house such incomplete, nonessential, and unfixed agents. Thus, those committed to democracy must ensure there is always space and time for a discussion of the arrogance of the center or dominant culture, especially a dominant culture that believes it is capable of including, representing, and adapting to others. As Young (1997b, 350) writes,

> If you think you already know how other people feel and judge because you have imaginatively represented their perspective to yourself, then you may not listen to their expression of their perspective very openly. If you think you can look at things from their point of view, then you may avoid the sometimes arduous and painful process in which they confront you with your prejudices, fantasies and understandings about them, which you have because of your point of view.

Such discussions cannot occur if the past is denied or bracketed out in order to create a necessary albeit false sphere of equality. If there are to be political public spheres that can address disquiet and offer space to perform and voice the concerns of those traditionally marginalized and excluded then there must be active listeners who strive to hear stories that are genuinely different from their own. As feral citizens have no home to lose and are perpetually changing, the riskiness of listening is replaced by a perspective that sees this risk as an opportunity to live, to hear, and to change. Feral citizens have no need to fear the interaction as they have no affiliation to that which could be threatened. In fact, feral citizens ask for the same kind of respectful and serious engagement when they visit communities that hold onto prejudices and beliefs that hinder the ability to hear or take seriously the challenge of a passionate and radical democrat.

Bell hooks, offers a hint at the kind of disruption that would accompany a core taking the challenges of peripheries seriously when she encourages the feminist movement to actually listen to those who have not felt welcome. Her guiding premise is that we "will know that white feminist activists have begun to confront racism in a serious and revolutionary manner when they are not simply acknowledging racism in the feminist movement or calling attention to personal prejudice but are actively struggling to resist racist oppression in our society" (hooks 1984, 55). Her challenge has the intention of exposing what she sees as oppressive elements of "sisterhood" feminism. Her radical and disruptive engagement with feminism is partially informed by her belief that feminism needs to be a part of a larger counterhegemonic movement that consistently disrupts the dangerous potential to digress into hegemonic conservatism. It also stems from a broader conversation that relates to antidemocratic, imperialist, or patronizing threats that continually develop around assumed symmetrical reciprocity.

Once again, the challenge to so-called common spaces is a reminder to those of us inhabiting "we" spaces that "many of the perspectives and practices that we take to be essentially constitutive and unquestionable aspects of our identity" (Coles in Bickford 1996, 150) are not beyond reproach. "Others, who explicitly or tacitly suggest that what we hold dear is trivial, illusory, oppressive, obnoxious, slave-like, unhealthy, and on and on" (Bickford 1996, 150), will always remind those in "we" spaces that when there is an "us" there is also a "them," which means an ideal democratic state is never achieved.[8] So for hooks, gender like any other identity should not be the permanent center or trump position of an agent with multiple subject positions and yet it remains an important position from—and to—which to speak. Culture, class, race, sexual orientation, location, age, and education must also be considered.

Audre Lorde (1997, 375) similarly warns of the danger of prioritizing particular identities when she explains that "those of us who stand outside that power [common we space] often identify one way in which we are different, and we assume that to be the primary cause of all oppression, forgetting other distortions around difference, some of which we ourselves may be practicing." Differences are part of the human condition and should never be permanently bracketed out for simplicity's sake, but nor must they be seen as a threat. They represent an opportunity to see whether the core has the capacity to listen to challenges that cannot be ignored by any movement that genuinely wishes to be part of the forces of freedom against the forces of tyranny.

As disruptive and challenging as hooks's critique is, it is certainly not suggesting feminism ought to be abandoned. Indeed, the struggle between respecting and celebrating individual freedom and uniqueness and achieving even temporary collective solidarity or equality constitutes a central component of contemporary feminism. Many feminists have taken challenges like hooks and Lorde's seriously and because they have accepted that what they have heard requires genuine changes they have made space for individuality and critique within their common—and now accepted as tension-filled—"we" spaces.

Rosi Braidotti (1994), for example, calls for diversity and tension within the feminist movement. Being a strong advocate of nonreciprocal responsibility and the abandonment of identity as a starting point for political action, she believes the developing political promise of what she calls difference feminism lies in its unique "epistemological position" (1994, 23) that consists of a combination of "coherence with mobility" and "combines features that are usually perceived as opposing" such as "a sense of identity that rests not on fixity but on contingency" (1994, 31). Her intention is to develop a way of "refiguring a subject position that is politically invested in the task of redefining his/her own accountability" (1994, 169), rendering both the agent and her interlocutors active listeners and courageous speakers. As with hooks, Braidotti sees feminism as a part of a political project that must embrace the tensions between collective equality and individual freedom. For Braidotti, the history of feminist tensions between homogenization around the collective "we" and fragmentation around the independent "I" are indicative of, and have many lessons to share with, those interested in the broader condition of globalization the contemporary Western world faces.

The uniqueness and particular relevance of Braidotti's theory lies in her intent to describe what she calls "the new nomadism of our historical condition" (1994, 169) and to "legitimize feminist theory as both critical and creative" by "reinventing a new kind of theoretical style, based on nomadism" (1994, 37). Nomadism is, according to Braidotti, a shift in consciousness that encourages one to live creatively and find liberatory promise in a condition that is already present and especially present among women. Nomadism is about inhabiting positions of power and changing them. Like feral citizenship, nomadism and embraced nomadic consciousness offers a way of politicizing and finding promise in a personal and political condition of fragmentation and multiple differences.

While itinerant,

> the nomad does not stand for homelessness, or compulsive displacement; it is rather a figuration[9] for the kind of subject who has relinquished all idea, desire, or nostalgia for fixity. This figuration expresses the desire for an identity made of transitions, successive shifts, and coordinated changes, without and against an essential unity...the point of being an intellectual nomad is about crossing boundaries, about the act of going, regardless of the destination. (Braidotti 1994, 22–23)

The "political agency" that emerges from nomadic consciousness, Braidotti (1994, 35) explains, "has to do with the capacity to expose the illusion of ontological foundations." Her political agent, again like the feral citizen, is focused on disrupting comforts and repoliticizing the political, which she describes as the activity that emerges from the "awareness of the fractured, intrinsically power-based constitution of the subject and the active quest for possibilities of resistance to hegemonic formations" (1994, 35). To be political means to resist hegemonic formations and rather than act as expected to perform through and as resistance to such formations. Thus, through political agency the already present nomadic consciousness of women becomes a contributing force for democratic change as it takes the actuality of women's lives and uses this actuality to critique the status quo and create or offer alternatives to the foundational assumptions of modern liberal democratic politics as well as much modern feminist politics.

Braidotti's theory of nomadic consciousness is right in emphasizing the importance of wandering and disruption to a politicized feminism; where it falters slightly is in not fully acknowledging the particularity of the nomadic subject she celebrates. All women can be nomads but surely, some have a greater opportunity to act on their nomadic opportunities than others do. It seems likely that not all feminists can or even should take on the privileged status of the nomad. "The nomad," Braidotti (1994, 4) explains, "is my own figuration of a situated, postmodern, culturally differentiated understanding of the subject in general and of the feminist subject in particular." I think the significance of the "my own" statement is more important than Braidotti acknowledges as not all women claim the privilege of being playfully or seriously nomadic and not all women would find being nomadic a privilege. Austin's walking woman, for example, was certainly nomadic but I dare say she would not have seen this as a privilege as much as a need or uncontrollable drive. It is for similar concerns that feral citizenship is never defended as an inclusive and necessarily liberatory solution to citizenship activity.

Even with these limitations and the recognition that Braidotti's nomadic theory does not "get it right," her attempt to "think through and move across established categories and levels of experience: blurring boundaries without burning bridges" (1994, 33) remains a democratically inspiring way of responding to critiques of sisterhood without denying the temporary relevance of the solidarity constructed through it. Nomadic consciousness requires each of us to be aware of our location(s) and take responsibility for the spaces we occupy. Those with privilege must challenge the privilege while taking advantage of it to act.

Everyone may have the right to have rights and everyone may have the right to be nomadic but the opportunity to act on these rights and opportunities is not equal. For Arendt, rights are rights of opportunity: the right to act and the right to speak. These rights are never equal and it is only once that is realized that "we" spaces can accept their promises and limits. For Braidotti, rights of opportunity are accompanied with obligations to use those opportunities accordingly.

In all cases it comes back to seeing oneself in relation to the broader community, acting without expecting reciprocity, and acting on behalf of a better future. Nomadic consciousness for its part represents an example of how one might attempt to straddle the line between universality and difference, and it represents a way of using knowledge of the irreducibility of difference as a part of an epistemological and political strategy. Both hooks and Braidotti ask women to abandon identity in order to allow for the construction of subjectivity. Subjectivity only ever appears within relationships and is thus performed, multiple, and necessarily collective. This focus on subjectivity is another way of focusing on the who of a person rather than the what. However, Braidotti is not denying the relevance of the what. In fact, those with rights of opportunity are the ones imbued with responsibilities that emerge from identities that are undeniable and should not be ignored.

While hooks's critique of sisterhood and Braidotti's nomadism uncover an interesting legacy of disturbance and difference within feminist theory in general, the most influential and disruptive Western feminist theorist to fully embrace and theorize the political relevance of border thinking, trespassing, and hybridization has to be Donna Haraway. In the next section, I look at how Haraway encounters the struggle between individual freedom and collective equality and suggest that her methodology and her disruptive activity represent another politically significant example of how one might perform as a flawed agent within always-limited public space.

An Active Listener's Response to Taking Others' Stories Seriously

Haraway's (1991) careful and playful retort to an impossible request to write a "keyword" entry for a Marxist dictionary represents one kind of mischievous activity that stems from demands placed on those who genuinely attempt to listen actively to others who are "for somewhere else and for something else." It is also a useful example of the tensions that those struggling to defend and expand difference face when confronted with important tasks that appear to contradict such ideals of difference.

The term Haraway was asked to define was "gender."[10] Reflecting on the fact that she was entirely incapable of defining the term, she came to realize that

> the evidence is building of a need for a theory of "difference" whose geometries, paradigms, and logics break out of binaries, dialectics, and nature/culture models of any kind. Otherwise, threes will always reduce to twos, which quickly become lonely ones in the vanguard. And no one learns to count to four. These things matter politically. (1991, 129)

They most certainly do matter politically!

What makes Haraway's contribution to the dictionary so extraordinary in both content and style is that as conscious as she is of her ensuing failure to actually offer a reasonable definition of "gender," she does not use it as a reason to refuse the request. She wrote the entry but she wrote it knowing full well that it would not achieve the desired or prescribed intention. Haraway used the opportunity to reverse the power relation by challenging the assumptions of those who gave her the invitation. The invited refused the conditions of entry but nevertheless took the opportunity of the invitation to enter on her own terms in order to perform as who she is rather than what the core wanted her to be (female Marxist).

Rather than simply giving up on the problematic request, Haraway struggled to come up with an appropriate way of participating; and by doing so she herself better understood the limitation of language. In fact, by attempting the project, Haraway was able to make it evident that *no one* could "justifiably" write such an entry if the purpose was to give a clear definition of what gender meant. The danger, she recognized, was that the desire to simplify obscures the often constitutive friction that has made a term publicly relevant and in need of (re)definition. To define is to fix, but as Haraway's participation shows, sometimes to define *unexpectedly* can be to challenge the assumed simplicity of defining and knowing terms that do not lend themselves to simple definition. So while a definitional project is never neutral, the universalizing rules of its project need not be followed. Haraway played the game, but played by her own rules informed by both humility and pleasure in the opportunity or privilege to play.

Haraway's defense of situated knowledge is similarly exemplary of a politically disruptive approach to inevitably flawed action in volatile times. Situated knowledges "require that the object of knowledge be pictured as an actor and agent, not a screen or a ground or a resource, never finally as slave to the master that closes off the dialectic in his unique agency and authorship of 'objective' knowledge" (Haraway 1991, 198). Haraway's noninnocent and nonidentarian approach to political participation is explicitly disruptive and implies an irreducible need for space to exchange opinions and stories.[11] Assuming the absence of a master story to tell (and no master storyteller), as Haraway does, offers the potential for each story to inspire, disturb, and create. New relations appear while old ones are destroyed and/or reformed. Situated knowledges do not guarantee, but certainly offer the possibility of creating an epistemological project that allows active listening and genuine difference to be constitutive components of epistemology and politics.

To help show how knowledge is always situated, relational, and conditional and to remain mysterious herself, Haraway, throughout her work, repeatedly reinvents herself

> as other by using invented figures as her native informants. In *Primate Visions*, apes are not simply objects of research, but coauthors of the tales. In *Modest_Witness*, her questions begin from the starting point of OncoMouse™, a corporate-engineered mouse used in genetic research. In "Otherworldly Conversations," she explores the instrumentality of languages from the dog's-eye view. (Bartsch, DiPalma, and Sells 2001, 139)[12]

By practicing this unusual tactic, Haraway disrupts the dichotomy between subject and object making her objects of research "collaborators, actants in the process." This process echoes Lorraine Code's "healthy skepticism" toward knowledge creation, in which she argues that we "shift epistemological inquiry away from autonomy-obsession toward an analysis explicitly cognizant of the fact that every cognitive act takes place at a point of intersection of innumerable relations, events, circumstances, and histories that make the knower and the known who they are, at that time" (Code 1991, 269). "Taking subjectivity into account," as Code (1993, 36) once again points out, "does not *entail* abandoning objectivity," it entails accepting the relational (not relative) nature of *both* objectivity and subjectivity. A rejection of objectivity as the trump of subjectivity does not require subjectivity to be "all there is"; it simply requires seeing more than the given, more than the obvious, more than the represented, and more than the stereotype. It also offers a way to allow for active listening.

Bartsch, DiPalma, and Sells (2001, 132) describe situated knowledges as "a dynamic and fluctuating comparison of not only marginal positions to the dominant one, but marginal positions to each other as constructed through vectors of power." Situated knowledge is relational rather than relative and, like "gender," it can be read as a claim for the irreducibility of difference and a defense of continuous political activity because it is always being reinvented as a result of new situations and new relations.

There are different ways of responding to Haraway's and Code's epistemological challenges. One is to try to achieve greater truthfulness, all the while knowing it is partial and situated as "translation is always interpretive, critical, and partial" (Haraway 1991, 195). Another is to admit that, given the absence of generalizable truths and complete or inclusive "we" spaces, we must situate our discussions in a sphere of opinion, a political sphere where relational and partial standpoints are respected, debated, and further developed and changed. It is the latter of these two responses that feral citizenship encourages.

Feral Citizens as Political Ramblers

Feral citizens are no more or less intrigued by oddities as they are by norms. Those seen as acting "acceptably" are just as interesting as those vilified by their unorthodox acts and positions. Feral citizens may mimic and adapt but will not *necessarily* do either. Being free from a place to call home and speak and act authoritatively from means there is no authenticity to be hidden, thus feral citizens invite anyone to perform and care very little for manners or rules of conduct. Furthermore, they are comfortable entering other spaces as performers, spectators, and guests regardless of the presumed rules of conduct and regardless of whether they have been invited or not. As opposed to being border creatures, they are creatures who ignore borders altogether.

Feral citizenship is a response to the absence of common space and the gradual loss of the Arendtian "in-between," so it encourages gathering together for storytelling and performance to help re-create political space and common worlds. Common worlds emerge when a plurality of people become bound by traditions, stories, and rituals that they share. In today's world, individuals have loose affiliations with many such worlds; feral citizens help solidify the loose affiliations and encourage their performance within new spaces that can in turn initiate new common worlds. However, for acting feral citizens creating moments is a never-ending practice.

When it comes to sitting at the table—which is the metaphor Arendt uses to describe the common world that both brings the public together and separates all those who have come to the table, the feral citizen will likely remain standing. Feral citizens want to be a part of the world "that is common to all of us and distinguished from our privately owned place in it" (Arendt 1958, 48), but for Arendt the common world is created by those who live together. As she explains, the common world or table "is what we have in common not only with those who live with us, but also with those who were here before us and with those who will come after us" (1958, 50). In other words, it is spatially limited and requires a set of traditions and beliefs that precondition the possibility of the world appearing. Feral citizens resist any such preconditions and refuse these kinds of conditions on their activities.

In any case, the spatial limitation may not be as important to Arendt as it first appears to be as Arendt (2005, 106) also suggests that "whenever people come together the world thrusts itself between them, and it is in this in-between space that all human affairs are conducted." Thus, in the modern less spatially limited world it is easy to envision the tables as the we spaces spoken of earlier, and it is easy to imagine individuals sitting at a number of different tables. Yet even in times where people are part of numerous worlds, each individual remains limited by the undeniable fact that the world only ever reveals itself to individuals from particular perspectives. It is this final point that helps encourage feral citizens to continue on their journeys. Arendt's (2005, 168) statement that the engaged citizen should look upon the world with "the greatest possible overview of all the possible standpoints and viewpoints from which an issue can be seen and judged" seems to suggest she would encourage such continuous travels within contemporary society. But for Arendt visiting had always been done with the intention of attaining greater truthfulness when considering an issue that needs judgment. Feral citizenship is not about increased truthfulness, rather it is about offering an additional perspective from which to view and engage in the world. The perspective of the feral citizen is not necessarily more truthful or somehow better than any other perspective; what it is, is different, and this is where its primary political promise lies. What it does is offer a new perspective which adds to the plurality of viewpoints which is a valuable political act in and of itself.

Arendt's well-traveled citizen is an improved citizen that can take on the role of a more honest representative of the public. Feral citizens are less about being better than they are about being other, remaining other, and being valued as other. Their contribution is as an addition and not a replacement, which is why feral citizens do not want followers—they want listeners, interlocutors, and diverse others who all have unique contributions to make in their own unique way. Returning to a point made in the previous chapter, this new political actor fits particularly well with the theater metaphor that Arendt uses to describe the kind of ephemeral activity that occurs in the in-between spaces that exist once people gather together in the common world. Unlike other kinds of art, the pleasure of the theatrical performance is exhausted in the moment. It is entirely reliant upon relationships and different viewpoints from the actors, to the audience, to the stagehands and all others. The result of the performance is not some kind of fabricated product like in painting or sculpture but rather a moment to speak or re-create with others that were present and others that were not.

Explaining the uniqueness of the performative art to politics, Arendt (1958, 209) explains,

> While the strength of the production process is entirely absorbed in and exhausted by the end product, the strength of the action process is never exhausted in a single deed but, on the contrary, can grow while its consequences multiply...the reason why we are never able to foretell with certainty the outcome and end of any action is simply that action has no end. The process of a single deed can quite literally endure throughout time until mankind itself has come to an end.

Performance "is the only art whose sole subject is man in his relationship to others" and only through performance is "the political sphere of human life transposed into art" (Arendt 1958, 167). The fact that performances offer no material residue means both the performers and the performance can only be judged relative to "the criterion of greatness because it is in its nature to break through the commonly accepted and reach into extraordinary, where whatever is true in common and everyday life no longer applies because everything that exists is unique and *sui generis*" (Arendt 1958, 184). Feral citizens offer a disruptive addition to the theater as both actors and audiences. They are less trained or interested in being trained and as a member of the audience they may well be far more active than the presumed silent audience. In other words, the feral citizen deprofessionalized and deformalizes the theater encouraging more of a guerrilla theater than the formal theater with a clear division between the performing professionals and an attentive, privileged, and respectful audience.

By deprofessionalizing performance and moving the theater from the stage to the street, citizen actors perform whenever the situation, relation, or moment calls for it. Theater becomes participatory and spontaneous, while performance becomes open to all rather than only those with a particular skill and/or virtue.[13] Prioritizing the political allows the roaming theater to become the potential political stage of the active and engaged citizen. Replacing the virtuosity and glory that accompany formal theater with desire, passion, and adventurous spirit that accompany street theater allows amateur citizenship to involve the creation of public theater in the streets.

To perform well and honestly, political actors need to be freed from the need to follow strict scripts or act according to some sort of external standard; they must be free to perform, and by doing so further create and nurture their desires whenever and wherever they choose. Performances then, become like the enunciative moments of which Homi Bhabha (1994) speaks; they create in-between spaces among the actors, the storytellers, and the audience, and they have the potential to re-create borders to allow for constant change and altering of relations, thoughts, and subject positions.

If feral citizens are perpetual wanderers who tell stories as and about political performances, it is reasonable to assume that the feral citizen will rarely guide the discussions or choose the performance to take place. Her or his role is to help make sure the natal and thus disruptive moments are able to become, or be temporarily translated into, part of a broad never-complete counterhegemonic methodology or politics of disruption. While particular purposes of the disruptive acts are certainly relevant within themselves, it is the collective legacy of disruption that renders the acts so important to feral citizenship. Feral citizens take seriously Bickford's (1996, 86) warning that at Arendt's table "each person has his or her own chair (and so particular location)," which means opinions and relations within the public sphere develop relative to their place at the table. Even if the table is round "as seems appropriate for the world" (Bickford 1996, 86), particular relations develop as a result of who is closest to whom and who guides and initiates the conversations. These relations can become predictable, sometimes stagnant and, as we saw in hooks's resistance to sisterhood, rarely conscious of the power skews they perpetuate. Feral citizens, thus, also take on the particular task of encouraging movement at and around the table while also reminding those present that the table is actually a table among many tables that collectively make up the public sphere. Ideally, feral citizens and acts of feral citizenship represent nonauthoritative stimulants to performance and discussion. They make the best out of the situation they are faced with; they are necessarily interactive and only temporarily present; they do not determine the content of the interaction; and they fully acknowledge that the conversation and performances continue (and are possibly much more dynamic and interesting) when they are not present.

This chapter has not given a clear description of what the method of a feral citizen would look like. But then that was never the intention. Rather, the intent was to hint at the enticing political promise of feral citizenship and the related need for activities like those encouraged by feral citizens. Postcolonial theorists and difference feminists have created challenges to epistemology that require similar developments in the political arena if they are to realize their disruptive-democratic promise. I have shown that feral citizenship and the creation of informal micropolitical moments offers ways to resituate Arendtian political practices in the present. I have also explained that feral citizenship is the sort of disruptive political methodology that acts on and contributes to the liberatory potentials opened up by the critical insights of those who recognize the promise of foundation-free thinking and acting.

As an explicitly antiauthoritarian methodology, feral citizenship deprofessionalizes and extends the sphere of politics by instigating opportunities to perform, watch, and tell stories outside (and not necessarily in relation to) formal political spheres. The situational inescapability of each of these moments is precisely what makes the persistent revitalization, disruption, and subsequent creation of political acts and deeds so essential to democratic society.

Difference feminists and postcolonial theorists have long been defending variance and marginality while finding creative promise in terms like consensual dissensus, situated knowledge, and relationality. Like feral citizenship, these challenges to knowledge construction presuppose a strong commitment to the democratic ethics of freedom and equality. The commitment, I have shown, may not be explicit. Indeed, many of the theorists I have referred to throughout the chapter resist any affiliation with political and/or democratic theory. Nevertheless, their critiques and their disruptive engagement with political and democratic theory, along with their clear recognition of the irreducibility of difference, make their work as relevant to the democratic theory they often challenge as it is to the places they call home.

To conclude we can say that as a method of political engagement, feral citizenship is a type of synagonistic political strategy by which I mean feral citizenship is a type of activity that is based on and encourages respectful struggle among those who have a shared enemy (Karagiannis and Wagner 2005, 241). For feral citizens, the common enemy is authority, the loss of the public sphere, and the reduction of politics to administration. The respectful struggle is (or ought to be) between feral citizens and practitioners of social movements and political theories and theorists struggling to resist the always-present threat to critical democratic activity. Feral citizenship is a way of attaching a synagonistic political theory to a method of political agency.

While feral citizens are purposely ambivalent, when it comes to their relations to order, there is an important relation to the foundational democratic culture that emerges from the current political condition within pluralist Western democracies. Feral citizenship is a particular response to particular times. The terrain to trespass on and into is conditional on modern circumstances. Synagonism is a theory of the present; feral citizens are actors of the present. Feral citizens create foundations when they instigate political moments and act on their democratic imperative. The imperative is likewise conditional and relational to the common democratic ethic shared with the agonists who are part of the struggle against the loss of political space and discourse. As will become apparent in the rest of the book, the tension or paradox between any foundation and a fixed (anti)foundational ethic is constitutive of many disruptive acts initiated by feral citizens.

Chapter 5

Public Realm Theory, from State to State of Being/Becoming

Public realm theorists have been debating the purpose, procedures, and activities of democratic public space since the Athenians' first short-lived foray into politics more than 2,400 years ago. While this chapter does hope to contribute a few new tensions and insights into this debate, its primary task is to draw attention to the shared commitment of all public realm theorists to divide the responsibilities of the public sphere into distinct realms with each having its own particular rules, tasks, and purposes. The motivation behind this particular intention comes from my belief that the mutual and foundational distinction all public realm theorists' make between the kinds of activities that are appropriate for different spheres has yet to be given the credence it deserves. Many of the most promising contemporary social movements, not least of which is the environmental movement, continue to struggle with reconciling the need to fight for survival along side the equally essential passion to help create a world where to be an environmentalist (or any other progressive) would not involve a constant battle against societal norms.

Some Preliminary Observations: Let One Hundred Flowers Bloom, Let One Hundred Schools of Thought Contend

One of the first things to note about contemporary discussions around liberatory politics is the success[1] radical democratic scholars have had in convincing a largely Leftist audience that liberal democracy is something worth struggling for. The aim of these liberal radical democrats has been to radicalize, pluralize, and/or rationalize liberal democracy in a time more definable by fragmentation than unity, more by privately based individualism than collective political beliefs, and more by passive acceptance of institutionalized authority than active resistance to corrupt relations of power.

Yet, radical democrats are not a homogenous group of scholars. While they appear to share an optimistic belief in the potential of liberal theory, the democratic political system, and current social conditions, they differ in how they wish to harness this potential. There are those who, in spite of the absence of a coercion-free space, a consensus-based public realm, and a common space of appearance (Villa 1992), continue to search for rationalistic conceptions of the public sphere. With the ideal of indicating the minimal conditions necessary for a public realm free from internal and external coercion, these theorists search for ways of legitimizing and pushing the limits of present-day democratic institutions through public participation and collective adherence to accepted norms. The most notable theorist behind this approach is Jürgen Habermas, who has struggled throughout his career to reclaim both the epistemological and political aspects of a modernist project that he believes was initiated by the rise of the bourgeoisie within Europe in the seventeenth and eighteenth century.[2] His particular political interest is with finding appropriate procedures that will allow pluralist society to reach rational and legitimate decisions on issues of public import. While aware that the ideal speech situation necessary in order to achieve legitimacy for decisions made may not be realizable, he and other deliberative democrats nevertheless believe the attempt remains both necessary and sufficient for rendering actually existing democracies legitimate.[3]

There are also those within the radical democratic tradition, such as Chantal Mouffe and Ernesto Laclau, who in light of the current conditions of fragmentation, diversity, and complexity believe the goal of radical democracy is not to find ways of legitimizing failing institutions but rather to find ways of harnessing the democratic promise of the antagonism and agonism inherent to a pluralist society freed from the tribunals of history and the metanarratives typically associated with such tribunals. Contextualists like Mouffe and Laclau believe understanding the exceptionality of the present is essential if the perpetually unfinished project of the democratic revolution is to continue to create liberatory opportunities in times distinctly different from any previously experienced. One might reasonably suggest this way of viewing the present is necessary if the democratic promise of the numerous occupy movements (occupies), the Arab Spring, the Quebec student tuition battles, and the many antiausterity struggles are to be understood.

It was nearly 30 years ago that Ernesto Laclau and Chantal Mouffe first argued that the task of the Left is not to "renounce liberal democratic ideology but on the contrary, to deepen and expand it in the direction of a radical and plural democracy" (1985, 176). Unlike Habermas, however, they do not strive for consensus or even legitimacy in/of decisions made. Nor do they believe—as Habermas does—it possible to hold onto the rationalist and epistemological elements of liberalism if one is to be capable of thinking about politics in today's complex pluralist times. Rather, as suggested above, they argue that if the unfinished project of the democratic revolution is to be realized only the political aspect of the enlightenment tradition can be resituated in contemporary democratic society (Mouffe 1995, 259).[4]

Premised on the belief that traditional democracy cannot deliver on its promises of collective equality, individual freedom, and inclusive civic participation, the above mentioned deliberative and antagonistic approaches represent two distinct and original attempts to answer the question of how to situate democracy and reclaim (the) public sphere(s) in spaces radically different from those in which democracy was born. Each endeavors to transgress the typical communitarian-liberal dualism in order to come up with a radically democratic revitalization of the public sphere in a time that they agree is defined by a nearly universal allegiance to democracy, an unprecedented proliferation of social movements, and an absence of any universally accepted metanarrative or authority.[5] Each offers ways of rekindling debates around political concerns, primarily related to how to reclaim and revitalize the political public sphere. Their common struggle is not with saving liberal democracy *as it is*. Rather, the guiding desire is to allow liberal democratic societies to develop into what they *ought to be* if they wish to stay true to the ethicopolitical ideals of individual freedom and collective equality. So while their theories and positions are not the same, radical democrats do take for granted certain allegiances to liberal democracy.

The unfortunate result of the assumed finality of liberal democracy has been the exclusion of more radical and/or republican inspired attempts at the revitalization of the public sphere from their conversation. The omission is primarily due to the fact that republican theories tend to have a much more critical view regarding the promise of liberal democracy whether it is so-called radical or not. In this chapter, Hannah Arendt and Cornelius Castoriadis, the two most notable Western (republican) political theorists excluded from this not-so-consensual conversation, are reintroduced as essential contributors to contemporary public realm theory. As well as helping to expand the boundary of what counts as acceptable political questioning and discourse, the inclusion of Arendt and Castoriadis helps to bring back a more genuinely radical, critical, and liberatory tradition of democratic theory largely absent from most liberal-inspired "radical" democratic theory.[6] Their presence also reintroduces the notion of politics as a space of freedom and public debate rather than merely, as Mouffe (2005, 9) would have it, "a space of power, conflict and antagonism," or as Habermas (1996b) would have it, a set of procedural norms that can achieve acceptable decisions within diverse and conflictual societies.

If, as suggested above, Mouffe's agonistic politics offers a lens through which to view the promise of modern-day democratic eruptions such as the occupies, the Arab Spring, the Quebec tuition battles, and antiausterity struggles across Europe and the United States, the manner in which many of these eruptions have organized, dwelled in, and taken public space both temporally and materially are best understood through the lens of republican theorists like Arendt and Castoriadis. This is primarily because the beauty of the movements is that they are less movement than they are events or rather many events that collectively represent an unprecedented coming together of both physical and ideological proportions.

As Arendt and Castoriadis are more critical of liberal democracy than Habermas or Mouffe, the primary effect of their inclusion into radical democratic discourse is to instigate a necessary challenge to the current allegiance to liberalism however weak it may be. The presence of Arendt and Castoriadis means that claims that liberal democracy is the best we can do (Habermas 1996a, 382) or that the task of the Left can no longer involve renouncing liberal democratic ideology (Laclau and Mouffe 1985, 176), become subject to discussion, as does the effect of leaving more radical arguments outside the parameters of the discourse as it defines itself.

In what follows I use Habermas as both a focal point and an entry into contemporary public realm theory. I chose Habermas for two main reasons. First, given the prolific nature of Habermas's contribution to public realm theory one cannot reasonably dismiss his influence on the renewed interest in the public sphere.[7] Second, I believe Habermas's elaboration of the distinction between weak and strong public spheres creates an opportunity for respectful debate between public realm theorists who have genuinely different perspectives as far as the purpose of the public sphere is concerned. As mentioned earlier, Habermas is not alone in drawing attention to different spheres and responsibilities within the public sphere. Indeed, Habermas's particular distinction between weak and strong public spheres is emblematic of an important commonality among public realm theorists that is rarely given the attention it deserves by Habermasians or his critics. He is also probably the most influential and certainly the most reformist of the public realm theorists discussed in this chapter so he represents a reasonable and common entrance into the tension-filled terrain of public realm theory.

Weak and Strong Public Spheres[8]

In the following section, I maintain that Habermas's weak public sphere can be both a legitimizing and a delegitimizing tool for present-day democracies. This dual possibility gives it an unpredictable potential that could conceivably lead to unimaginable processes and outcomes including, but certainly not limited to, rational opinion formation and legitimation. So the following section goes beyond what Habermas wants to teach in order to examine what Habermas might offer once his particular intention of politics as legitimation tool is opened up to other possibilities.

What Habermas distinguishes between is a "constitutionally protected" creative sphere oriented toward opinion formation and a more formal administrative public sphere oriented toward responding to and acting on the formed opinion. The former is creative and free due to the absence of the need to decide, while the latter is more structured and limited due to its role as translator of the generated public opinion into a demand that can be turned into law or legislation.

Habermas's (1996a) recognition of the need for creativity in the public sphere has always been present in his work; however, it has rarely been a focal point. This is because for Habermas, the creativity within the weak public sphere is relevant as a means to an end and not as a possible end in itself. In other words, its relevance is directly related to its capacity to perform an initiating role along a chain of events moving from creativity to public opinion to legislation, law, or policy.

There are two ways of participating in the Habermasian weak public sphere. The first is through the signal function, which "acts as a warning system with sensors that, though unspecialized, are sensitive throughout society" (Habermas 1996a, 359). The role of the participant here is to communicate or uncover problems that can then be "processed by the political system." The other prescribed role of the weak public sphere is the job of "effective problematization" of the signaled issues. For effective problematization, the participant is to "convincingly and influentially thematize" issues of public import. The way to thematize is to furnish publicly relevant issues "with possible solutions, and dramatize them in such a way that they are taken up and dealt with by parliamentary complexes" (Habermas 1996a, 359). Both of these tasks are oriented toward influencing the state and providing an acceptable illustration of public opinion.

Yet if participants within the weak public sphere are to be seen as legitimate representatives of public opinion, the public sphere must be open to the spontaneous development of "open and inclusive network[s] of overlapping, subcultural publics having fluid temporal, social, and substantive boundaries" (Habermas 1996a, 307). Therefore, a functioning weak public sphere must remain free from external forces of any sort and must always remain open to the concerns of those who may not typically represent the majority of the public. The necessary openness means the weak public sphere is also "a 'wild' complex that resists organization as a whole" and is thus always hard to translate and use for means beyond its own engagement with those who are present in the sphere. Habermas is suspicious of this necessity and is concerned with the weak public sphere's vulnerability to nondemocratic or imposing actors. In other words, he is worried that the very nature of the sphere (lack of preexisting structures) makes it susceptible to unequal participation and all sorts of dangerous antidemocratic rhetoric and action (Habermas 1996a, 307–8).

Habermas is correct to be concerned but as with so many others who are suspicious of the freedom, spontaneity, and unpredictable nature of democratic association he responds by imposing an instrumental filter onto a noninstrumental space. What is needed is a celebration of the danger and an acceptance of the threat as an essential part of a healthy democracy as it is in these informal spaces that actually existing democracy is found and then shown to be wanting. To be wild is to be free and as Thoreau correctly states *in wildness is the preservation of the world*. Furthermore, a vibrant weak public sphere is also a critical weak public sphere and the threat of exclusionary politics and tactics cannot be discussed or addressed anywhere better than within an engaged radically democratic public sphere oriented toward performing and being an example of freedom, equality, and social solidarity in action.

If a vibrant weak public sphere is both creative and critical there is no reason to limit the value of such a public sphere to Habermas's legitimacy role. In fact, there is good reason not to. The Arab spring, the maple spring, and the international occupies are all cases in point. Many of these creative and nonstructural uprisings are not interested in informing the state, convincing the state, or creating public opinion. The demands, desires, and actions of these movements transgress the capacity of the state to respond and are the most promising enactments of democracy we have seen certainly since the early days of the anti- or alter-globalization movements.

In many of the uprisings, the audience has not been the state but the state responded in order to try to attain its legitimacy and relevance (reminiscent of the state response to the Ramblers). The reaction of the state due in part to the incapacity of the state and its foot soldiers to comprehend that which it cannot translate, indicates how relevant it is to consider the weak public sphere as much more than a legitimacy tool. Occupies may have drawn attention to the lack of legitimacy and there are cases of reasonable responses from certain cities such as Oakland, New York, Boulder, and Albany but this is not where its political promise lies. As anyone who has been following the uprises knows the promise lies in its creativity, its lived expression of democracy, and its performance *in* the realm of appearance but not *of* the realm of appearance. Occupies challenge those who try to understand it and speak for it, and encourages engagement with their messages rather than consumption of their stories.

The far more radically disruptive promise of the "constitutionally protected" weak public sphere is presently also being realized in the antiausterity uprisings around the world; each time the state attempts to tame these uprisings, its incapacity to do so becomes more and more evident. Occupies have performed moments and created events that are not translatable into legislation or policy. The weak is becoming strong but the capacity to understand this lies in freeing oneself from precisely what Habermas is trying to chain us to—the legitimacy of the state. Perhaps, the language of the 99 percent has helped change public opinion and perhaps the passion of the occupiers has gained more sympathy for those who have always occupied or dwelled within the weak public sphere, but one thing is clear and that is the fact that the democratic promise of occupies lies outside the state–civil society relationship.

The long overdue expansion of the weak public sphere and the immeasurable creativity occurring within many places and spaces that are contributing to this expansion are not primarily focused on the state, they are not making demands, and they are not acting in a predictable, indeed reconcilable manner. This lack of performing the prescribed and easily understandable role of the Habermasian weak public sphere, of course, is why they were being criticized by the media and government for not having leaders or a clear platform. Their actions were and are creative and far from fitting into the parameters of society, they challenge them. Not surprisingly mainstream media is not only incapable of translating the message but it is also incapable of hearing the message. There are reasons both in and of the media for this incapacity but as with the lack of focus on the state most occupies have not got caught up in the attempt to translate an untranslatable message into a space that does not lend itself to creativity or genuine challenge. Occupiers are rightly not interested in the container offered to those participating in the mass media spectacle. Their avenues for sharing information and stories are informal ones not structured and controlled by those who have their own message to protect and enhance.

If occupies, antiausterity battles, and the democratic eruptions surrounding the Arab spring can be viewed as performances within the informal public sphere (I cannot see any other way of interpreting them), it seems like contemporary weak public spheres have more in line with the Arendtian (1963, 23) concept of public life where "the polis was supposed to be an isonomy, not a democracy" and where people gather together for joy, resistance, and natality not responsibility and collective legitimacy.

Actors within these movements and moments are also not only making arguments but they are also introducing new ways of being present within political spaces that are not only oriented to asking for handouts from, or making demands of, the state or other authorities. Through their actions they are showing that a specific type of discourse can never permanently embody the public sphere as certain individuals, language games, traditions, and identities will always emerge and show it to be flawed as far as achieving its legitimacy.

While I have been focusing on what I believe to be the democratic potential of the weak public sphere, I have also been careful not to ignore the fact that caging the potential of the sphere's wildness, while unfortunate, may be necessary if it is to fit with any idea of deliberative authority. Habermas's approach to a discursively rational political system requires a diversity of procedures that allow for participation in some part of the decision making process. By outlining or uncovering a plurality of spaces for participation, Habermas indicates the extent to which the voices of the people can be heard. By opening up political participation to numerous types of discourse, he succeeds in extending what political participation can mean, but each type of participation remains, in typical Habermasian fashion, limited to its role within the liberal democratic system. That is, the purpose of participation is translated into a legitimizing tool for the political function situated above it in the hierarchical political system.

Nothing in Habermas's political theory, not even the wild public sphere free from authority and domination has a political purpose in and of itself and this is his primary blind spot. The weak public sphere is, for Habermas, always related to the formal public sphere, which is far less encouraging and exciting than the weak public sphere.

To conclude this section, we can say that Habermas's split is a welcome one but the absence of reflection concerning the potential tensions between the spheres of activity means Habermas fails to take seriously the most promising developments within the informal public sphere. In fact, this failure represents one of the most notable limitations to the Habermasian split; concerns raised in public, if they are to be politically relevant, must fit within, accept, and be oriented to the purpose of the larger political system. They must be translatable into a form of public opinion that can get the administration to act on the deliberations of the public sphere. Stated slightly differently, particular issues must be translated into universally relevant moral issues that can be rationally deliberated. So while the freedom offered in this space is recognized as a necessary part of a legitimate political system, its political promise is limited by an a priori assumption that the final intent of the public sphere weak or strong is to achieve rational moral consensus.

The Formal Public Sphere, Rational Not Free

At first, Habermas's celebration of radical autonomy, followed by a caging of the communicative power of this wild public sphere, seems confused. When read closely in the context of his larger political goals, however, it makes "rational" sense to have informal opinion formation regulated by democratic procedures. Perhaps here it is useful to recall that Habermas has always been less interested in freedom and equality for the individual than in rational deliberation that can lead to reasonable decisions. With this focus, the need for distinct (informal or formal) bodies that work with one another becomes clearer. Strong public spheres with regulated democratic procedures are intended to take rational and universalizable conclusions and translate them into laws or policies. Yet even if Habermas is primarily interested in the legislative end point, his focus on procedures requires questioning how the consensus came about and how it is performed or presented;

these questions always lead back to the weak public sphere. So while there is danger in reducing the weak public sphere to a tool for public opinion, it is equally imperative not to misinterpret the consequence of Habermas's focus on procedures. He may be a reformer, but he is not oblivious to the claims of those who demand radical changes.

Habermas finds in the structured parliamentary public sphere opportunity to institutionalize procedures that can ensure decisions made are as legitimate and representative as possible. Reliance on parliamentary structures is where Habermas's reformism becomes most evident. It is also where he reiterates one of his most relevant points. There is a distinct difference between making decisions and reaching understanding. Decision making is a *need*. It stops the wild explorations of the weak public sphere and requires numerous limitations on individual freedom if it is to lead to decisions that can be considered legitimate and rational.

The task of making decisions is an administrative or necessary duty. It is a type of management necessary for dealing with the wild aspects of a complex society when politics is seen as responsible for achieving consensus. It requires behaving accordingly, not acting unpredictably, and while it may support laws that develop equality it also ought to be recognized as a hindrance to the democratic freedom of an active and pluralist citizenry.[9] Strong public spheres, more centralized and much less diverse than more exclusionary weak public spheres, are "arranged" prior to the need to decide and have particular rules that are to be followed for "justifying the selection of a problem and the choice among competing proposals for solving it." The structures are organized around the "cooperative solution of practical questions, including the negotiation of fair compromises." Their role is not to become "sensitive to new ways of looking at problems" but to respond to, justify, and evaluate the selection of the particular problem (Habermas 1996a, 307).

The parliamentary strong public sphere has to react to public opinion developed out of the weak public sphere by convincing the administrative sphere to institutionalize or legislate the final decision. The assumption is that the sole goal of political action is law. Indeed, Habermas's deliberative process is concerned with the question "what is *valid* law? Or, more precisely, how is a *legitimate* law, which necessarily involves a claim to transcendent *validity*, possible in a post-metaphysical context?" (Palti 1998, 118, emphases original). The way to achieve this, Habermas answers, is to ensure the "normative expectation of rational outcome is grounded ultimately in the interplay between institutionally structured political will-formation and spontaneous, unsubverted circuits of communication in a public sphere that is not programmed to reach decisions and thus is not organized" (1999, 57).

The role of the strong parliamentary public sphere is to ensure universality. It is also intended to institutionalize rational decision making, to be inclusive, and to be democratic. The need for a creative and just public sphere is also, without question, shown to be essential for a democratic society. The weak public sphere will, however, remain necessary until all irrationalities are eliminated, and as such a notion is itself irrational and highly utopian (dystopian?); it can be assumed that the weak public sphere will always be necessary and desirable. So Habermas's own rationality paradox means politics and undisciplined interaction remain essential parts of deliberative democracy. It is in relation to this sort of challenge that the other three public realm theorists in this chapter have the most to offer.

Chantal Mouffe (1993, 8), for example, has long argued that "for radical and plural democracy, the belief that a final resolution of conflicts is eventually possible, even if envisaged as an asymptotic approach to the regulative ideal of a free and unconstrained communication, as in Habermas, far from providing the necessary horizon of the democratic project, is something that puts it at risk." For Mouffe, the way to take advantage of modern pluralist conditions is first to accept the dimension of undecidability that infiltrates every decision-making moment. This undecidability is brought about by the irreducibility and contextual nature of all antagonisms, what Mouffe (2005, 17) calls "the hegemonic nature of every kind of social order and the fact that every society is the product of a series of practices attempting to establish order in a context of contingency."[10] Second, the political sphere is to become a means for temporarily translating antagonistic relations into agonistic relations. "While antagonism is a we-they relation in which the two sides are enemies who do not share any common ground," Mouffe (2005, 20) writes, "agonism is a we/they relation where the conflicting parties, although acknowledging that there is no rational solution to their conflict, nevertheless recognize the legitimacy of their opponents." This irreducible tension is why a vibrant public sphere is so important to agonistic pluralists like Mouffe. The idea is not to destroy the opponent but to engage the opponent as an agonist to persuade and also potentially to learn from. When a subject position of a citizen is threatened within this democratic and creative container the we-them relationship that is created initiates a political relationship and makes visible the relevance of the primacy of politics within a pluralist society. If the conflict is public, this political moment either demands response from political institutions or draws attention to its failures to achieve freedom, equality, and social justice. If the latter is the case, the political sphere where such agonistic relations emerge radicalizes and creates the kinds of connections that can transfer a particular issue into a broader political concern that may in turn reinvigorate the political and expand the political sphere.

Part of Mouffe's broader political intent is to allow particular issues or nodal points to take advantage of the plurality of these respectful disagreements among a diverse population of individuals with multiple subject positions in order to gather different democratic struggles together to temporarily form chains of equivalence that can become part of a new collective will or conflictual consensus. For Mouffe, it is this tension-filled collective will that constitutes a new temporary "we" space (or perhaps many of them) made up of numerous radical democratic forces that will inevitably find actually existing democracy wanting.

Here we see similarities to the discussions in the previous chapters especially in relation to the contingent, unfixed, exclusionary, and unfinished nature of all "we" spaces. We also see the relationship of the particular to the universal that stimulated hooks's friendly critique of sisterhood feminism. As hooks had no interest in destroying the "we" space created by feminism she was most certainly an agonist and the result of her engagement did lead to the realization of certain exclusionary and thus antidemocratic tendencies within the feminist movement.

There are also similarities to Arendt's celebration of action, which requires a beginning, a realm of appearance, a subject of the action, and a freeing of responsibility regarding what happens to that action once the natal point has lost its vigor. Nodal points, natal points, and acts or performances are unpredictable, spontaneous, and wild moments—consequences or even purposes are never knowable in advance but once they have occurred there is no going back. The act as a performance, as a relationship built between actors, between actors and audience, and between actors and future storytellers cannot be undone. Thus, the weak public sphere where such creative and dangerous acts develop is where citizenship is formed and performed. It is also a realm of freedom where one's nonrepresentable individuality is experienced and created. Its political value is ephemeral but the traces that are left on individuals and society are lasting and irreplaceable. To be understood, these moments must not be subject to a rationality filter or pragmatic evaluation. On the contrary, their primary political value lies in their capacity to create the conditions that can allow the incompleteness of each individual, each community, and each world to be experienced, seen, and acknowledged. They should be seen as miraculous, creative, and at their best, wonderful. As such, the way to understand and judge them is through the lens of greatness and the way to see such acts as great is to view them in relation to the political promise of each human within a democratic society that accepts democracy not as something one has but rather as something always on the frontier, something always incomplete, something inconceivable, indeed something great.

As incompleteness, diversity, and tension are the foundation and purpose of Mouffe's politics, it is no wonder that she is keen on moving the focus of radical democracy away from Habermas's all too easy add on to mainstream liberal democracy. However, Mouffe's vision of politics is unique and sometimes difficult to follow, so it is important to take the time to understand precisely what it is that distinguishes agonistic pluralism from other democratic approaches to the condition of plurality. Mouffe is clearest on this front when she describes the differences between what she calls "politics" and "the political." She explains,

> By "the political," I mean the dimension of antagonism which I take to be constitutive of human societies, while by "politics" I mean the set of practices and institutions through which order is created, organizing human coexistence in the context of conflictuality provided by the political. (2005, 8)

Further clarifying the importance of the separation (and showing her similarity to Castoriadis's instituting-instituted distinction, Arendt's politics-violence distinction and Habermas's weak-strong distinction), she explains that "it is only when we acknowledge this dimension of 'the political' and understand that 'politics' consists in domesticating hostility, only in trying to defuse the potential antagonism that exists in human relations, that we can pose the fundamental question for democratic politics" (1999, 754). The fundamental question of democratic politics, she argues, is not one of how to arrive at an inclusive rational consensus, but rather, how to create "unity in a [never reducible] context of conflict and diversity" (1999, 754).[11] The realm of politics for an agonistic model of democracy is thus oriented not toward consensus but toward creating new political frontiers[12] where agonists can engage with each other and create political spaces needed to challenge and constitute politics. It is thus transformation from competitive antagonism between enemies, to a respectful agonism between worthy adversaries that politics is to strive for in a fragmented, pluralistic, and particularized society. In this agonistic view, politics only ever temporarily slows down the political.

Reintroducing the Republicans

Hannah Arendt and Cornelius Castoriadis share Mouffe's passion for plurality but their more republican inspired politics means their interest is less about agonism than it is about natality, imagination,[13] and "the capacity to bring about the emergence of what is not given—not derivable, by means of a combinatory or in some other way—starting from the given" (Castoriadis 1997b, 104). For them, as with Mouffe, everything is incomplete and rather than solve this incompleteness, politics harnesses it. Politics is *to-be*, so political theory like Habermas's that rests on the belief that what exists is good enough, is in their view a threat to politics and in clear need of the imagination of autonomous and active political agents such as those dwelling in the weak public sphere.[14]

Castoriadis explains that for him the radical imaginary that allows for creative and innovative actions to emerge operates on both an individual and a social level. Individually, it acts as a source of irritation, creativity, and disturbance to the instituted world. Socially, the imaginary creates "social imaginary significations" that (un)consciously organize the meaning world of a certain society. The social imaginary is what helps individuals make sense of the world they are embedded in, but as stable as this imaginary appears to be, its apparent stability is always challenged by the creative, imaginative, and disruptive "instituting society" that houses a radical imaginary never entirely absent from any society. The imaginary not only informs, relies on, but also transgresses, instituted society and exists only within weak public spheres that far from serving the legitimation needs of the strong public sphere directly challenge and threaten its role within a truly democratic society.

Castoriadis is the most oppositional and explicitly radical of the public realm theorists included in this chapter and like Arendt, he is interested in action, freedom, and natality. The rationale behind his oppositional starting point is best summed up by his claim that "on the level of the real functioning of society, the 'power of the people' serves as a screen for the power of money, techno-science, party and State bureaucracies, and the media. On the level of individuals, a new closure is in the process of being established, which takes the form of a generalized conformism. It is my claim that we are living in the most conformist phase in modern history" (Castoriadis 1997a, 346). So while Mouffe sees promise in the translation of antagonisms into agonisms, Castoriadis would rather the direction be reversed with agonistic relations being realized as, or transformed into, oppositional relationships that are not about respectful disagreement but rather about actually destroying the other. The other, for Castoriadis, is modern-day instituted society.

Castoriadis argues that in the present society although individuals and groups are permitted to struggle for justice within particular parameters they are also kept separate and "repressed by the entire contemporary social structure, by the reigning ideology, by the tireless effort of the traditional organizations to suppress it [the germinal critique of the status quo], and, of course, by individuals' psychical internalization of this structure, the self-repression of new significations they create without completely knowing it" (1997a, 9). This is why he and other passionate celebrants of politics' democratic promise will never accept the reduction of the public sphere to the largely administrative tasks of deliberation, decision making, and proceduralism. It is also why more critical and genuinely radical democrats like Castoriadis must not be excluded from discussions concerning the democratic promise of the public sphere as they are willing to go beyond Mouffe's examination of friendly enemies within agonistic relations in order to explore the radically destructive and oppositional promise of the weak public sphere that may itself not only be populated by agonist but is also often driven into action by a common unfriendly enemy. In fact, while Castoriadis may share Mouffe's passion for plurality, he would be no less critical of her legitimizing and taming agonistic pluralism than Habermas's conformist and rationalist consensus. In an ideal democratic condition where autonomous individuals engage each other as equals with imaginative will and promise, agonism may be the appropriate way to view potential tensions and disagreements. But as, according to Castoriadis, we are very far from such an ideal state we must also reinvigorate oppositional battles and remind ourselves that some opponents and antagonists should not be translated into agonists. Sometimes, opponents are opponents for good reasons.

Castoriadis's focus on creativity and radical challenge to societal norms means politics must include "elucidating the problematic of revolution, of denouncing false-hoods and mystifications, of spreading just and justifiable ideas" (1997a, 33). This is clearly not about reducing antagonisms into agonisms; it is about radical critique and creative revolutionary (read oppositional) response to pseudodemocracy. "Democratic creation," for Castoriadis (1997a, 343), is "the creation of unlimited interrogation in all domains, what is the true what is the false, what is the just and the unjust, what is the good and what is the evil, what is the beautiful and the ugly." Such questions are genuine questions[15] that demand all institutions and norms are at least potentially subject to critical interrogation.

True politics, and the activities of the public sphere, is thus about creativity, the imaginary, and "the reorganization and reorientation of society by means of the autonomous action of individuals" (Castoriadis 1987, 77). Politics begins with "the explicit acknowledgement of the open character of its object and exists only to the extent that it acknowledges this" (1987, 89) and accepts that an autonomous individual's freedom is directly related to the freedom of others. Thus, politics is a never-ending movement striving to make society "as free and just as possible." It resists authority on all fronts and constantly disrupts instituted society through its creative capacity to introduce something previously unthought of.

Arendt's similarly antiliberal and antiadaptive politics is embedded in her explanation of the differences between the three activities of the *vita activa*: labor, work, and action. While she considers all three activities essential to humanity and "intimately connected with the most general condition of human existence, birth and death, natality and mortality" (1958, 10), it is only action that is distinctly human and distinctly political. It is also only action that requires a genuinely free weak public sphere. Very briefly, as Arendt has been discussed throughout the book, she explains that labor involves the business of keeping oneself and others alive, while work consists of the construction of artifacts that will transcend the limits of human mortality and provide permanence. These two activities, while important for securing the *survival* of the species, do little to promote the life or potential of the individual and for Arendt, should be kept out of political space. Rather, the individuality and distinctness of humanity depends on action as it has "the closest connection to the human condition of natality...the capacity of beginning something anew" (Arendt 1958, 11). For Arendt, to act and differentiate oneself from others is what distinguishes humans from other animals and it is through action that the promise of the human condition is realized.

For Arendt, the public sphere is a space of action—it is a place where individuals perform who they are or would like to be not what they are or have to be. The action that occurs in the public sphere is the only activity that goes on directly between men without the intermediary of things or matter, corresponds to the human condition of plurality, to the fact that men, not Man, live on the earth and inhabit the world..., this plurality is specifically the condition—not only the *conditio sine qua non*, but the *conditio per quam*—of all political life. (Arendt 1958, 9–10)

As was mentioned in the previous chapter "performative actions," which Arendt believes ought to constitute the activities of the political public sphere, "are not alternative ways of deliberating, rather they are agonistic expressions of what cannot be captured by deliberative rationality" (Kulynych 1997, 345). Like the radical challenges performed by Castoriadis's autonomous actors, they are the actions of distinct, unique individuals who are nonrepresentable and nontranslatable. Actions are about individuality and each human's capacity to stand out in a crowd and introduce the unimaginable. So, once again, they represent far more of a threat to the formal public sphere than a prop to it.

Arendt does have a formal or instrumental side to her politics but it is about ensuring the protection of unstructured and undisciplined public space for its own purpose, as it is while acting that men are free rather than possessing the gift of freedom (1968, 153).

For Arendt, political life is to include

the joy and the gratification that arise out of being in company with our peers, out of acting together and appearing in public, out of asserting ourselves into the world by word and deed, thus acquiring and sustaining our personal identity and beginning something entirely new. (1968, 263)

From the above, we can conclude that Castoriadis, Arendt, Habermas, and Mouffe each has a particularly important contribution to make to public realm theory, and as such there are lessons to learn from all four theorists. Part of my intention for including this chapter is to reinvigorate the debates and uncover the similarities—without obscuring the differences—between these public realm theorists. I do not, however, believe that by reinvigorating these debates an inclusive sphere of public debate will immediately, if ever, be created. Rather, more humbly, I believe that by acknowledging the different focus of each theorist I can make it far more difficult to justify the permanent dismissal of any public realm theorist when discussing the revitalization of the public sphere.

Politics, Feral Citizenship, and the Environmental Movement

Any approach to theorizing democracy must "make room for dissent and for the institutions through which it can be manifested" (Mouffe 1999, 756), but so too does it need to make room for cooperation. In much the same way as Mouffe and Habermas take from both the liberal and the communitarian traditions in order to engage in the messiness of the present, I want to use certain insights from both the deliberative and the agonistic approaches to assist in my own contribution to the revitalization of the public sphere. But I am unwilling to allow the differences between Mouffe and Habermas to contain the discussion concerning the revitalization of the public sphere in modern times. As I have shown, both Castoriadis and Arendt have much to offer especially in relation to their deeper critique of liberal democracy. However, I have at no time proposed bringing all these theorists together under one cooperative political umbrella. Each theorist has particular interests and desires that have both hindered and assisted their political theories and none should be sacrificed to a new superior hybrid theory. This is not the same as saying they ought to remain fixed in their positions and not listen to the arguments of others. On the contrary, it is to say listening to others must always be a part of any approach to revitalizing the public sphere. Any notion of "getting it right" should be resisted, and any opportunity to create spaces where differences can be performed, shared with others, and developed should be expanded. This is of course, what feral citizenship is concerned with. The idea, as I have asserted throughout the book, is not to replace what is already present but to create moments where other political promises can be understood, explained, disrupted, and expanded.

Briefly summing up these different approaches to politics, we see that for Arendt politics is primarily a joyful act that allows active and independent agents to gather together and perform the unimaginable while realizing their unique human capacities to act and be free. Politics, action, and freedom not only rely on each other but are also indistinguishable, as it is only when free from the sphere of necessity that the human capacity to act is unleashed, and it is only human action that can introduce the unexpected. Or, as Castoriadis might similarly argue, it is only the autonomous individual who can imagine an autonomous society. For Castoriadis, politics is about the activity of creating and the attempt to create autonomous individuals and societies. Mouffe, however, sees politics as directly linked to antagonism. She believes undecidability and antagonism are irreducible aspects of the social symbolic makeup of modern pluralist society and that the purpose of politics is to create conditions where the other is no longer viewed as an enemy but as a respected adversary to debate with and learn from. Last, Habermas sees politics as a procedural movement from the creation of public opinion to its institutionalization and legislation. Each approach has its place and time. Each ought to be understood as a necessary though not sufficient part of a continued attempt to revitalize the democratic public sphere. And the place of these necessary debates is the weak public sphere.

In this chapter, by arguing with, for, and against the political promise of Habermas, Mouffe, Arendt, and Castoriadis, I have attempted to return a favor granted me by each of these theorists. At one time or another, all four disturbed particular beliefs that I held prior to engaging with their work and each has influenced the development of feral citizenship as a joyful adventure through the wilds of democratic theory and practice. So rather than simply describe how each theorist has been helpful in the development of the concept of feral citizenship, I have used feral citizenship to return the disruptive favor by stimulating, further politicizing, and reinvigorating discourse between attempts to revitalize the public sphere.

The same lack of recognition or consensus over what politics, political agency, or citizenship means or even ought to mean haunts many other politically inspiring developments including those coalescing around the notion of ecological citizenship. In fact, the ongoing discussions around the purpose, promise, and danger of ecological citizenship in relation to environmentalism as a creative counterhegemonic movement represents an ideal example of the irreducible need for noninstrumental and engaging conversations. In the following two chapters, I argue that ecological citizenship's political promise lies in its capacity to create space for such disagreements and misunderstandings to be housed and discussed. As such, I argue against the tendency to filter out the complexities and tensions within the discussions and I argue against those who wish to claim the term for their own instrumental desires or to add legitimacy to their prepolitical intentions. In its place, I offer a more limited role for ecological citizenship as one small yet essential part of the environmental movement, and I suggest that if ecological citizenship is to retain its political promise it must be freed from the need to be the savior of the fledgling environmental movement.

Ecological citizenship represents a significant opportunity to reradicalize the environmental movement, but to realize this opportunity will not be easy. The specters of pragmatism, crisis, and necessity are always waiting to pounce on any political development within the environmental movement; this is why it is essential that such developments are seen as limited and only ever part of a much bigger battle that environmentalists know all too well. My proposed limited and humble role for ecological citizenship within the environmental movement is not shared by most other environmentalists interested in the articulation. This is because most turn to citizenship for legitimacy sake or for a means to act on, in, or against instituted society. I do, however, see room for such arguments within the broader intentions of many theorists who have taken the time to genuinely think through the articulation of environmentalism with democratic citizenship.

CHAPTER 6

A TOUGH WALK: ENVIRONMENTALISTS ON DEMOCRATIC TERRAIN

> *We do not boast that we possess absolute truth; on the contrary, we believe that social truth is not a fixed quality, good for all times, universally applicable or determinable in advance...Our solutions always leave the door open to different and, one hopes better solutions.*
>
> —Malatesta 1965[1921], 269

In 1992, Robyn Eckersley argued that one could discern three "major ecopolitical preoccupations" in green political thought. These were participation, survival, and emancipation; respectively, they correspond loosely to the three previous decades of environmental politics. In the two decades since Eckersley's book, there has been a new focus among green political theorists, which could be called the *democratic* preoccupation. While a great deal of variety remains within this stage, almost all take a human-centered[1] view (Jelin 2000, 49), and almost all attempt an articulation between ecological or environmental thought and democratic theory. Specifically, green political theorists have, as John Barry (1996; 1999, 193) points out, realized "that any plausible modern political theory embodies a commitment to the view of individuals as deserving of equal respect and concern," suggesting democracy must be "an essential part of all political theories," including environmental ones. Green citizenship,[2] a particularly intriguing subsection of this "new" stage, is the focus of this chapter and the next.

One often hears that "a good citizen is one who obeys the laws, pays taxes, votes ritualistically for pre-selected candidates, and minds his or her own business" (Bookchin 1987, 10). If this is true, it is not in being a "good citizen" that the promise of green citizenship lies. Rather, it is to be found in its potential to disrupt both radical democratic theory and green political thought; it lies in its creation of and participation in green public spheres that are themselves situated in broader democratic-political public spheres; and it lies in its potential to once again render environmentalism a radical challenge to the status quo both in mainstream and radical discussions within green political thought and democratic theory.

Green citizenship resituates environmentalism in a new discourse and a new time, and similar to the feral citizen, its presence and the politicization of nature that it brings to political communities can be very disruptive. Yet too, from an environmental perspective, it can be very risky if the purpose is truly to situate representations of nature within political discourse and render it subject to political rules of the game. There are no guarantees that this new democratic stage will better protect the environment or lead to more sustainable societies. There are not even any guarantees that the articulation of green citizenship will assist in the extension of democratic politics. A turn to politics means a turn away from guarantees. It requires taking risks and it means situating environmentalism in a much broader and never-ending struggle between freedom and domination. So, again, like the feral citizen, the green citizen has limits as well as political promise and its political promise relies on it not being asked to do more than it is capable of as a political actor. Likewise, it relies on its limits being accepted as essential aspects of citizenship rather than inconveniences to ignore or transgress once it is resituated in environmental politics.

Douglas Torgerson's book *The Promise of Green Politics, Environmentalism and the Public Sphere* (1999), and his contribution to *Environmental Politics*, "Farewell to the Green Movement? Political Action and the Green Public Sphere" (2000), are emblematic of this democratic stage and succeed in giving a good overview of the sort of tensions that one might expect would guide ecopolitical discourse within the democratic stage. In fact, his convincing argument that "green politics poses a challenge to the instrumentalism of the industrial world, throwing into question the hierarchical and often technocratic tendencies of modern governance while promising a new kind of politics" (Torgerson 2000, 1), is one of the clearest introductions to the latent democratic promise of green politics, a promise he argues that is *already* present. I share Torgerson's belief that there is a largely unrecognized potential in the ideals of much green political theory. I also applaud his turn to Hannah Arendt, as opposed to more common invocations of Jürgen Habermas and other proponents of the deliberative approach to pluralist conditions. Further, I agree with Torgerson (2000, 1) that while green political thought does offer challenges to present-day instrumentalism, it also has a strong tendency to replace many of the challenges with its own equally limiting instrumentalism.

A collection of essays in *Environmental Politics* that claims to represent "one of the foundational statements in what will surely become a lengthy conversation" (Dobson and Saiz 2005, 157) about the relevance of citizenship in an environmental context, seems to attest to the danger Torgerson speaks of. Dobson and Saiz, the authors of the introduction and editors of the issue, explain that there is a notable "consensus among analysts of this turn to citizenship that the very enlisting of the idea implies a recognition that sustainability requires shifts in attitudes at a deep level" (Dobson and Saiz 2005, 157), going well beyond fiscal measures or policy initiatives.[3] I concur that shifts at a deep level are necessary and that there is a great deal of political promise in the articulation of green political thought with citizenship. I also share the authors' belief that "citizenship has come to seem important in the environmental context" (Dobson and Saiz 2005, 157) and that the consequence and stimulus of its relevance is deserving of "lengthy conversation." But I notice a clearly instrumental and purposeful undertone to the turn to citizenship, and I am unwilling to accept the editors' finding of a conclusive consensus, as it comes dangerously close to closing off or drastically limiting the discursive space or "lengthy conversation" that a turn to citizenship requires now, and will continue to require if it is to remain politically promising.

Discursive spaces opened up by interactions between theories of polis citizenship and theories of green citizenship offer an unprecedented opportunity for directly challenging and disturbing the most common hindrances to creative discourse within and between both schools of thought. It is this intention to bring together two cultures that are yet to be fully acquainted that van Steenbergen (1994) was interested in when he first spoke of ecological citizenship and it is this intention that Dobson (2003) himself is most interested in his book *Citizenship and the Environment*. Unfortunately, this potential remains obscured by the current dominance of instrumentalism and end-focused intention by most green political theorists who have made the turn to citizenship. As difficult as it may be, the current attempt to close these conversations down or limit their ability to learn from each other must be resisted by green citizens and for green citizenship if its political promise is to begin to be realized.

Teena Gabrielson (2008) has also critiqued the instrumental focus of green citizenship claiming it narrows the promise of citizenship, but unlike my attempt to limit the realm of citizenship in order to resist the loss of citizenship's promise her intent is to expand the realm of citizenship to include noninstrumental aspects of political ecology. Early on in her article Gabrielson refers to van Steenbergen who was one of the first to speak about ecological citizenship. Van Steenbergen wanted to create a conversation between two cultures who he correctly believed had yet to be fully acquainted. His intention was to encourage a creative interaction where the outcome between two cultures lamenting their lost glory days could produce new imaginaries that might transgress the limits of both cultures. Gabrielson suggests that such a conversation has been hindered by a narrowness within both cultures but especially the environmentalist one. I agree with Gabrielson's observation that this meeting has yet to produce the political promise van Steenbergen hoped for, but I also believe while Gabrialson is correct to challenge the narrowness of the conversation due to the instrument baggage attached to the environmentalist culture the response ought not to be an expansion but a limiting of the range of possible articulations and creative moments that might emerge from this distinct, creative, imaginative and temporary engagement between two cultures that need to be disrupted but not destroyed. The idea ought not to be for a kind of dialectical emergence where the limits of the past are transgressed in order to make way for a new and improved ecological politics that can become the new vanguard for the twentieth century. Rather the ideal ought to be the addition of a new conversation that adds to the necessary plurality of a polity that has thankfully chosen justice over a common good and thus operates within both cultures while concurrently transgressing the limits of each. It is this kind of approach that is offered by feral citizenship and it is this kind of approach that emphasizes creativity and change both at the individual level and at the cultural level and thus helps constitute the active and critical public sphere so important to a healthy democracy and healthy planet. So while Gabrielson's rationale is compelling, I worry that the proposed widening may have the same depoliticizing effect as the narrowing does. As I have mentioned previously, citizenship needs to be distinct if it is to retain its disruptive political promise. So we must never forget to ask the question why citizenship?

Gabrielson (2008) asks why citizenship and convincingly argues that citizenship must be given a greater emphasis in the partnership between green thought and political agency and she is without a doubt correct in suggesting such an emphasis will challenge the apparent (ab)use of citizenship as a way to legitimize predetermined green intention, but I do not see a widening of the discourse as the best response, at least not from a politics-first perspective. If anything, the best response may well be a more realistic expectation of the role ecological citizenship might play within the broad and diverse green public sphere. The problem is not in the narrowing of citizenship's promise but in the wrong kind of narrowing. Noninstrumental issues should not only be included in ecological citizenship's focus but also they should replace the instrumentalism that does and always will obscure its political promise. Assuming, as Gabrielson (2008) does, noninstrumental issues can exist alongside instrumental ones within a school of thought littered with crisis talk and apocalyptic scenario is an assumption that I am unwilling to make. It is simply too difficult to gain an audience and create meaningful political moments with those who speak in the language of crisis and apocalypse. If green citizens are to 'focus on the democratic and egalitarian aspects of citizenship in their theorizing (Gabrielson 2008, 430) it is only going to happen once crisis talk is excluded and political rules of engagement are followed.

If citizenship can be more than a legitimizing tool for environmentalists, there must be a willingness to move beyond what can only be seen as the greatest hindrance to green citizenship becoming a part of the democratic movement or the forces of freedom. That hindrance is the seemingly essential articulation of environmentalism with ecological truths. This partnership, certainly central to, and an important component of early environmental ethics or thought, is not a partnership that can hold once green citizenship is situated within radical democratic discourse. In fact, if it is true that "some strands of green thinking are actually inimical to the concept of citizenship" (Dean 2001, 490), the "partnership" between ecology and citizenship is surely a main one.[4]

The fact that the environmentalism-ecology partnership continues to haunt theories of green citizenship points to the fact that the true liberatory potential of the green-citizenship dialectic is yet to be realized. It also points to a very real danger in the articulation of green concerns with democratic citizenship when the green issue is viewed as prepolitical and the idea of citizenship or politics is taken for granted. It seems that far too many green political theorists have been influenced by Robert Goodin's (1992, 168) claim that "to advocate democracy is to advocate procedures, [while] to advocate environmentalism is to advocate substantive outcomes." This claim, while welcome as far as an implied challenge to the often-assumed easy partnership between democracy and environmentalism, has led much environmental-democratic discourse down a path that has obscured the most liberatory ideas of democratic theory and environmental politics.

As far as the latter is concerned, the minimalist view of politics as procedural decision making and environmental concerns as necessarily a particularly relevant (more important than other less urgent concerns) part of the procedure, has obscured an already present challenge to instrumentalism by promoting a "distinctly instrumentalist cast" of its own (Torgerson 2000, 1). The focus on substantive outcomes, rather than the more abstract allegiance to what Chantal Mouffe (1993, 51) calls the continuation of the democratic revolution, "understood as the end of a hierarchical type of society organized around a single substantive conception of the common good, grounded neither in Nature or in God," has led to a lack of theorizing concerning the potential and limit of participation in radically democratic societies that must exclude instrumentalism if they are to realize their political promise.

CITIZENSHIP AND THE PRIMACY OF THE POLITICAL

> While it is perfectly true that we cannot save humanity unless we save the earth, there is no purpose in saving the earth at humanity's expense...there is no reason to reject the possibility of human emancipation. (Dean 2001, 502)
>
> Citizenship is always in the process of construction and transformation. (Jelin 2000, 53)

As a commitment to democracy is the most recognizable common ground between traditional notions of citizenship and green citizenship, the first step in elaborating on the liberatory potential of this partnership is to create and make clear the common ground for creative conversation. By establishing, however loosely, a shared understanding of what democracy ought to be, it becomes possible to create conditions of engagement that allow for the discourse to be oriented not toward a desired end, but toward the creation of a green public sphere;[5] and, I would add, more general public spheres that neither inhibit nor necessarily prioritize the emergence of the concern for nonhuman nature.

As democracy is the necessary condition for the stimulation of meaningful discussion between agonistic pluralists and theorists of green citizenship, it is much more than a partner to either theory. It represents the irreducible and constitutive ground from which the opportunity for discourse emerges, making the commitment to democracy, not the commitment to sustainability, the essential binding component of the desired affinity. Namely, a substantial commitment to democracy represents the necessary condition for the realization of democratically honest green citizenship.

If citizenship is "a form of political identity that consists of an identification with the political principles of modern pluralist democracy, that is, the assertion of liberty and equality for all" (Mouffe 1993, 83), then committed citizens whether green or not must be reminded of their primary allegiance to democracy. In a highly pluralist and diverse society, citizenship is the one common identity that everyone shares, and while it is always incomplete and in need of disruption, so too must it be protected so as to remain common.

Citizenship only makes sense in the context of a public good and the public good is democracy or the acceptance of justice, freedom, and equality. It is this common good that feral citizenship consistently reminds all others of. When one speaks, acts, and listens as a citizen s/he speaks, acts, and listens as a situated and relational public agent concerned primarily with increasing the democratic promise of the political public sphere. The important thing for the green citizen, then, is to make sure representations of nature are not systematically excluded from the *possibility* of emerging and when they do emerge making sure they emerge as equals rather than add-ons or trumps. By ensuring green issues are not systematically excluded from political discourse and by temporarily representing the nonhuman within political discourse, green citizens can take advantage of the opportunity to translate environmental concerns into political concerns that can help in the expansion of the political public sphere. However, political rules of the game must be followed and it must be recognized that an honest entrance into political discourse, at least radically democratic political discourse, requires that environmentalists accept a number of difficult preconditions for involvement.

First and foremost, participation requires replacing expert-dependent discourse with a sort of *enviro-doxa* where claims to truth are replaced with more humble and open claims to worthy opinion that can be "revealed, contested and changed in the company of multiple others by way of a performance[,] the persuasiveness of which is judged in the realm of appearance itself (exp. aesthetics), and not with the reference to either inner authenticity or external standard" (Sandilands 2002, 3).[6] But realistically opinion, if green, is informed prior to the sort of political engagement being suggested here. The simple discussion of the potential assumes the preconstitutive nature of the desired interaction. Thus, while the ideal would be for the green concerns to emerge out of political discourse, the reality is that environmental issues are already part of public discourse and cannot be fully bracketed out of the discursive make-up of the public sphere whether it is desirable or not.

The fact that green concerns are already central to political discourse is one reason why the shared intent to blur the distinction between the public and private spheres of action by both radical democrats and theorists of green citizenship is so significant. While citizenship requires listening, interacting, and respectfully conversing, it need not deny or ignore the never completely absent history of the actors. Environmentalists must be open to hearing others but they need not abnegate themselves or their thoughts in order to do so. Furthermore, the promise of the disruptive nature of the environmentally inspired citizen already exists, meaning it is translation and resituation not obliteration that is required for environmentalists to take the turn to citizenship seriously.

The Public-Private Controversy

While the differences between Mouffe and Arendt have been largely glossed over in this book, there is one important difference that should not be ignored when it comes to exploring the political promise of green citizenship. This is the difference between the way the two theorists call for participation within, and wish to protect the vitality of, the common ground of plurality and the political sphere. For Mouffe the public-private distinction is temporal not essential as it is in Arendt.[7]

Mouffe is not alone in her desire to denaturalize and de-essentialize Arendt's public-private separation. In fact, it is a part of Arendt's argument that has been challenged, quite rightly, by numerous feminist political theorists (Pitkin 1981; Honig 1992; Fraser 1997; Curtis 1999). Of particular concern to many critics is Arendt's commitment to a Greek understanding of politics that rests on the notion that "the freedom of the political realm begins once necessities are dealt with" (1958, 118) and her equally troubling argument that "where life is at stake all action is by definition under the sway of necessity, and the proper realm to take care of life's necessities is the gigantic and still increasing sphere of social and economic life whose administration has overshadowed the political realm ever since the beginning of the modern age" (1968, 155).

Reliance on the Greek tradition is typical among republican thinkers and while certainly important as a germ (to borrow a phrase from Castoriadis), is also clearly limited as far as responding to modern conditions of plurality and extended participation. Liberal or libertarian ideals, which focus more on plurality, border crossing between private and public issues, and freedom from the majority or central authority need to be included as countermeasures to republican ideals of equality and common will.

Mouffe, with her keen focus on antiessentialism, argues not that the personal or private is political, but more reasonably, and more in line with Arendt's general distinction, that the personal or realm of necessity *can become* political. Mouffe suggests that nothing is inherently apolitical; neither is anything necessarily political, and nor does any issue need to remain politicized once it has been rendered political. Honig (1992, 222) makes a similar point when she suggests a performative and antiessential reading of Arendt's distinction between public and private spheres can free action from its own "home" in the political sphere. Her point is that action, as a "self surprising" moment, can appear and politicize those spaces typically free from its influence. As "there are numerous instances of permeation of these distinctions" (1992, 223) throughout Arendt's work, Honig suggests it is not unreasonable to believe a blurring of the boundaries can allow politics to enter into and influence nonpure political realms of activity. But how does this move allow the political gaze into apolitical spaces without turning everything into a transparent and open sphere of politics where everyone is an *equal* citizen?

It is in answering this question that the crucial significance of Mouffe's coemphasis on antiessentialism and nonfixity surfaces. What Mouffe offers is a way of allowing the creative sphere of politics to disturb the closed spheres of intimacy and necessity. Her theory of agonistic pluralism allows the citizen the opportunity to temporarily situate politics within the private sphere of the household in order to open up relations to the public gaze and make the translation from private "I" to public "one" or "we."

What is most important, and what Arendt seems not to consider, is the fact that the participating subject or subject position could return to the household whenever desired or needed. Politicization, for Mouffe, is not a permanent act and the translation from "I should not be treated this way" to "one should not be treated this way," is necessarily temporary in order to retain the individuality of the "I."

The key difference lies in the influential liberal side of Mouffe's argument that seems to better respect the otherness as well as the changeability of individuals. This kind of commitment to temporality and nonfixity is particularly relevant when considering the politicization of nature. The act of representing nature means rendering nature knowable, taking the unknown, wild, and wonderful, and making it appear knowable or at least representable. The act of representing nature, perhaps even more than other representations, obscures as much as it enlightens. This means that if we are going to politicize nature or speak for nature it is essential to ensure that "nature" does not become permanently situated in human discourse even if that discourse is radically democratic. Nature must also be able to return to the mysterious realm of the unknown after it has been politicized or figuratively brought into the public sphere through human representations.

The act of politicizing nature creates a paradox whereby nature must be rendered knowable in some manner in order to be protected as a wild, unknowable other. I have argued in this section that a way for theorists of green citizenship to approach this paradox is by making clear the commitment to nonfixity and antiessential discourse, a commitment that can be best defended once the concerns around representing nature find a place in political discourse. I have also argued that the political discourse must be willing to move beyond rigid distinctions such as those between public and private spheres and those between human and nonhuman needs.

We can turn to other critiques of Arendt's rigid distinction to help further accentuate and clarify the relevance of blurring these divisions. This time the best example comes from one of the strongest (and still best) critiques of Arendt offered by Hannah Pitkin (1981). Pitkin takes Arendt to task for failing to acknowledge the social conditions that lead to political interaction. She states, "Our public life is an empty form—at best a meaningless diversion for a few, at worst a hateful, hypocritical mask for privilege—unless it actively engages the unplanned drift and the private social power that shape peoples' lives" (Pitkin 1981, 346).[8] Expanding on this point, Pitkin explains that the job of the political actor or citizen is to make social questions, or those questions related to the realm of necessity, "amenable to human action and direction." By opening up the opportunity for politicization, Pitkin, like Honig earlier, refocuses Arendt's fixed or material boundaries by proposing the use of more symbolic boundaries whereby "the danger to public life comes not from letting the social question in, but from failing to transform it in political activity, letting it enter in the wrong 'spirit'" (Pitkin 1981, 346). This argument, like Mouffe's, does not deny the relevance of the Arendtian distinction, but simply points out the fact that Arendt's rigid distinction does more harm to her political argument than it does good. What Mouffe, Honig, and Pitkin succeed in doing is situating Arendt's argument in a modern context that is quite unlike the Greek polis Arendt is fond of referring to for justification of her political arguments.

The danger that Pitkin, Honig, and Mouffe point out is precisely what those wishing to come up with a theory of green citizenship face. Certainly, there are many reasonable arguments against situating environmental concerns or representations of nature in political communities. Such arguments are most convincing when environmentalists remain reliant on ecological truths, survival talk, and crisis rhetoric to support their positions. In each case, the "spirit" of entrance is unacceptable and the individual citizen responsible to "a set of political principles...or 'grammar' of political conduct" (Mouffe 1993, 65), is replaced by the collective need to be a behaving species[9] responsible to something larger such as survival or big wilderness. Such a view is particularly evident among those environmentalists who continue to support a stewardship ethic that requires a great deal of discipline and behavior modification and very little active citizenship (more on this later).

A way to approach the survival spirit in nature discourse, as Pitkin (1981, 348) points out in another context, is to focus on self-realization rather than self-interest or self-sacrifice. Pitkin's claim stems from the aforementioned translation of a private "I" to a public "we" where the particular is temporarily transferred into a more common concern such as "no one should be treated like this." That is, when one's opportunity for self-actualization is wrongly thwarted, the collective "we" if asked, can, and should step in. Of course, when it comes to the nonhuman, the argument that "no one should be treated like this" is integral, yet by no means simple. That nonhumans animals lack the ability to translate the "I" into a "we," and that no one (as in human) should be treated the way many nonhumans and their environments are treated, is commonly accepted. What is not commonly accepted is that *those* animals ought not to be treated as such or that claims for justice can or ought to be extended to the nonhuman. The idea of nonhuman animals being granted the condition to achieve self-actualization is most certainly a political idea, and while we may not know what those conditions are we can be quite sure about certain conditions that will hinder them (factory farms, zoos, laboratories, circuses, etc.). One of green citizenship's responsibilities must be to help explore the political implications associated with translating the human "we" into an interspecies "we." The translation is not needed to allow nonhuman animals to participate in political discourse. Rather, it is needed to allow human consideration for their well being and otherness access to political conversation.

Rebecca Solnit, in a beautiful passage at the end of her book on the history of walking introduces some of the kinds of environmentalist beliefs and influences that might reemerge once environmentalists reconsider their role as members of the political sphere. She explains,

> Musing takes place in a kind of meadowlands of the imagination, a part of the imagination that has not yet been plowed, developed, or put to any immediately practical use. Environmentalists are always arguing that those butterflies, those grasslands, those watershed woodlands, have an utterly necessary function in the grand scheme of things, even if they don't produce the market crop. The same is true of the meadowlands of imagination; time spent there is not work time, yet without that time the mind becomes sterile, dull, domesticated. The fight for free space—for wilderness and for public space—must be accompanied by a fight for free time to spend wandering in that space. (2000, 289)

By emphasizing antiessentialist conditions of participation, there are many limitations on green citizenship, limitations most actually existing theorists of green citizenship would be unwilling to accept. However, as part of the intention of this chapter is to see how and if green and democratic citizenship could coexist, acceptance by others at this time is not a determining concern.

In the next chapter, I explore the promise of particular approaches to green citizenship in relation to how they can assist in the expansion of the public sphere. The disruptive and wandering methods of feral citizenship, while not always explicit, are obviously central to the interrogation practices as is the belief that a diversity of approaches to citizenship, as long as they remain democratic, is much better than one approach that attempts to be inclusive.

CHAPTER 7

THE OBSCURED PROMISE OF GREEN CITIZENSHIP

The key differences between proponents of green citizenship lie in how they intend to "deal with," embrace, or situate ecological disruptions. By way of introduction to stewardship, care, and deliberation—the three main attempts at "dealing with" the tension—I begin this chapter with a short look at Andrew Dobson's contribution to the debate. Just like one could not discuss public realm theory without addressing Habermas one could not reasonably talk about ecological citizenship without giving Dobson his rightful place in its center—if indeed it has a center. While there is much to learn from Dobson's work, it is his desire to reconcile, and admitted difficulty in reconciling, an ecological with a democratic ethic that I find most promising as far as offering a humble starting point for ecological citizenship as a disruptive influence on both ecological and democratic theory.

DOBSON'S ECOLOGICAL CITIZENSHIP, PLENTY OF ECOLOGY, NOT SO MUCH CITIZENSHIP

> The first virtue of ecological citizenship is justice. More specifically ecological citizenship...aims at ensuring a just distribution of ecological space...[I]t is relations of systematic ecological injustice that give rise to the obligations of ecological citizenship. (Dobson 2003, 102)

In the introduction to his book *Citizenship and the Environment*, Dobson states that the book is intended as a "contribution to the debate about how to achieve a sustainable society" and he claims that "ecological citizenship" represents a potentially important and "underexplored" contribution to the promotion of such a sustainable society (2003, 4). Thus, Dobson is looking to ecological citizenship as a means to an end and he has a particular, if slightly myopic, focus when looking to or for promising approaches to citizenship that can house his intent. In fact, it is clearly his instrumental focus that attracts him to a cosmopolitan approach to citizenship as cosmopolitanism is associated with obligation, a common humanity, and a commitment to strangers. Such an association means duties are owed nonspecifically, they are nonreciprocal, they are nonterritorial, and they are horizontal rather than vertical as far as being between citizens rather than between citizens and the state (2003, 116).

His intention is to promote a postcosmopolitan approach to ecological citizenship that replaces the weak nonobligatory ties of cosmopolitanism with a thick community of historical obligation tied to one's ecological footprint. Added to the thin community of "common humanity" is the far thicker obligation that emerges from the political space constituted by the ecological impact of human activity within an unequal globalized world.

For Dobson, the principle virtue of ecological citizenship is justice (2003, 113) and to be just one must ensure their ecological footprint is sustainable rather than unsustainable (2003, 116). In this case, the citizen is obliged to act once it is shown that his/her footprint is too large and as a result s/he has caused harm to another by taking more than his/her share of the earth's resources. If I am the one doing harm I, as a postcosmopolitan ecological citizen, am obligated to act and reduce the harm. So it goes without saying Dobson is less interested in the political promise of the partnership and the ensuing tensions than I am.

Nevertheless, there is much to like about Dobson's contribution to ecological citizenship and I am largely on side with much of his project especially his willingness to see how disruptive ecological citizenship can be to traditional and contemporary notions of citizenship. However, as my intention is also to convince the environmental community to take the turn to citizenship more seriously, I have concerns with Dobson's reliance upon a prepolitical foundation to inspire citizenship action as well as his reluctance to consider the creative promise of citizenship due to his overreliance upon obligation and virtue attached to the seemingly irreproachable ecological footprint. One's ecological footprint may need to be a consideration in order to be a just citizen within a modern globalized world but surely the sphere of justice must go beyond a personal footprint, and the idea of ecological footprints as foundation for just acts must also be open to debate.

Tim Hayward (2006a) alludes to the limits of the ecological footprint as foundation when he challenges the absence of political space made available to those who are owed the obligation of the responsible postcosmopolitan citizen. Hayward (2006a, 439) asks, "How is it determined whether someone is an ecological citizen on Dobson's account?" He answers that it appears that "the ties of citizenship bind in one direction only, on the beneficiaries of the inequalities; the others are effectively cast in the role of 'moral patients.'" In other words, agency seems to be given only to those with a moral obligation to act. Dobson (2006, 449) responds to Hayward's criticism by reminding Hayward that his version of citizenship is about practice rather than status, and as such agency is determined by the way one answers the question: "Do I owe, or am I owed, ecological space?" As ecological citizens, those who owe are obliged to act and those owed are recipients of the justice done by the ower. I do not find Dobson's response particularly compelling even if the recipients of justice can also demand their rights and in any case there remains little room for the justice inspired relationship between owed and ower to expand beyond the ecological footprint container. In essence, those who are owed the obligation remain silent and external and those obligated are to be virtuous and burdened actors responding to truth claims made by those who calculate ecological footprints. Neither the oppressed nor the oppressors are seen as desiring agents, and both the oppressor and oppressed are subject to a new external force that grants them or takes away agency the moment they accept the new metanarrative of the ecological footprint.

No doubt the oppressed want and are owed their fair share of the earth's resources but perhaps they would like more or something other or as Hayward (2006a, 440) suggests perhaps they would like "a share of the benefits of ecological unsustainability." It seems clear to me, if, as Dobson suggests, justice is the key foundation for ecological citizenship it must move beyond the material focus of the ecological footprint and venture further into the realm of democratic justice that includes active listening and the possibility that the ecological footprint may not be the best foundation for the responsibility the globalizers have to the globalized. Justice is always complex, challenging and while the attempt to ground it in something tangible is understandable it might not be just, and it might not be the best way to achieve a more just, free, and democratic society that is central to cosmopolitanism and democracy.

Dobson's normative foundation (obligation attached to ecological footprint) leaves precious little space for voice or active listening on the part of the citizen. When a polis or political moment does emerge, it does so as a burden bestowed upon a guilty party that is to then take on this burden and act accordingly, virtuously, and responsibly. As Dobson (2006, 450) states "once we establish ecological space as the territory of ecological citizenship, and the ecological footprint as that which determines the obligation positions of people in respect to one another, we see that justice is the virtue currency of ecological citizenship." But once again, surely only those obliged to act need be virtuous and more importantly, where is the space for joy or even respectful agonism? Where is the space for creative resistance? Where is the space for agency and freedom? Where is the space to find links with other progressive movements not necessarily focused on ecological issues? Where is the space to challenge the antidemocratic nature of certain environmental foundations? These are the kinds of questions that emerge from my politics-first commitment to the partnership. They are less intended as a critique of Dobson than they are a challenge to his use of citizenship (especially as a practice) as the appropriate identity to attach his prepolitical obligation to. I think citizenship can and must be much more (and at times much less).

In one of his earliest essays on ecological citizenship, Dobson (1999) suggestively states that four "typical" binary oppositions are disturbed by the presence of ecological citizenship: the rights-duties opposition, disturbed by the nonreciprocating nature of ecological duties (1999, 6); the public-private divide, disturbed by the fact that "the ecological citizen operates at all levels of society" (1999, 10);[1] the active-passive divide, disturbed by the belief that the "citizen as a consumer is a very active individual" as s/he compares prices and demands satisfaction from public services (1999, 13); and finally, the territorialized-deterritorialized distinction, disturbed by the cosmopolitan focus of ecological citizenship (1999, 15). As welcome as these disruptions are, their remoralizing intent as well as their attachment to the ecological footprint means the disruptions are themselves nonreciprocal (as far as environmentalists learning little from entering the political sphere) and they are severely limited due to their containment in an apolitical belief that cannot help but encourage ecological citizens to impose their presence on others and limit rather than expand the political sphere.

It is certainly welcome to see that the disruptions require that ecological citizens perform their responsibilities without expecting equal treatment in return or increased rights as a reward due to "the unreciprocated and unilateral nature of the obligations of ecological citizenship" (Dobson 1999, 7), but the political sphere loses much of its value if these new participants cannot themselves also gain from the opportunity of coperforming in a political arena.

This quick glance over Dobson's contributions to ecological citizenship makes it quite clear that none of the disruptions he proposes is intended to democratize the political sphere or for that matter the ecological knowledge that informs ecological citizenship. Rather, normative duties, presumably informed by particularly enlightened ecological truth-tellers, are to become permanent parts of everyone's responsibility as dutiful earth citizens who are seemingly joyless and certainly not free and adventurous.[2]

The political potential of Dobson's vision is essentially bound by his intent to introduce preconstituted ecological truths to citizenship discourse. The disruption is clearly one directional and the truth of *the* ecological position, typically informed by exclusive spheres of knowledge like science or privately informed ethics and interests, are untouched. The purpose of political activity is no longer increased freedom but increased obligation and responsibility to predetermined goals. Citizenship becomes subject to a new imperative that requires obedience and dutiful activity rather than exploration and freedom.

If Dobson's theory of ecological citizenship is representative of the current state and intent of ecological citizenship, this new citizenship may well, as Dobson claims, be novel and disruptive, but its disruption leads neither to an expansion of the political sphere nor to a revitalization of democratic citizenship. Rather, its novelty lies in blurring boundaries in order to expand the sphere of individual duties and responsibilities that have little relation to any radically democratic practice of citizenship. Dobson's ecological citizen is a responsible and well-disciplined individual who lives a "proper way of life"[3] in public and private spheres. Add to this picture the absence of discursive space and of opportunities for undisciplined activity, and the idea of a world filled with ecological citizens becomes rather unpleasant (and certainly no more democratic than it is now).[4] So as useful as Dobson's book length conversation around ecological citizenship is, it is limited by his unwillingness to examine the promise of ecological citizenship as a contributor to the expansion of the political public sphere, which while it may not be the specific intention of ecological citizenship it ought to be its broad intention if it is not to fall into the ecorepublic-leviathan trap.

If citizenship is to remain a distinct political identity, the commitment to citizenship within green political thought needs to be taken much more seriously than Dobson does. Furthermore, if environmentalism's turn to citizenship is to assist in keeping citizenship distinctly political, green citizenship requires a strong push in the direction of democracy. It must take seriously both the limit and promise of citizenship and accept that, as with politics in general, "it is only by respecting its own borders that this realm [or practice], where we are free to act and to change, can remain intact, preserving its integrity and keeping its promises" (Arendt 1968, 264). Indeed, this idea of protecting the *particularity* of citizenship has yet to be fully considered. In the next section, I address this deficiency by highlighting a particularly democratic side of green citizenship.

Green Public Spheres and Democratic Green Citizenship

If the disruptions of which Dobson speaks are to help expand the public sphere or increase the sphere of democratic responsibility, they need to be situated in public spheres where they can be introduced to radical democratic ideals of freedom and equality. This noninstrumental potential of green citizenship relies on public spheres, where "the very process takes on value for those who participate in it" (Torgerson 1999, 129); in other words, where debate is recognized as intrinsically valuable and pleasurable. Green public spheres, interacting as a *polis*, create discourses that situate those theorizing green politics in relations that help them realize their own particularity. Importantly, the relation is to other green theorists, and if opened to a wider audience, to their own particularity in relation to broader democratic goals of increased freedom and equality. Theorists *of* the green political sphere need to also be theorists *in* the political sphere. The purpose of such green public spheres is not to achieve a desired end point such as sustainability but to understand others' positions and engage in creative conversation with them.

These green public spheres gain "credibility as a historical possibility for the simple reason that a discourse has emerged, making it possible to formulate and discuss ideals that industrial discourse formerly excluded or marginalized" (Torgerson 1999, 130). They take on the important role of uncovering much of what is obscured by dominant culture. So while an instrumental focus continues to influence the intent of most of those who propose an approach to environmental politics, Torgerson (2000, 1) nevertheless believes that there is a challenge to, and rethinking of, political action "already present in green politics." He also realizes that the challenge is often obscured in instrumental arguments like Dobson's that view politics primarily as a way of achieving the desires of the environmental movement. The danger for Torgerson is that without allowing the political side of environmentalism to appear within green political discourse an endorsement of the environmental position can easily lead to an endorsement of a new fundamentalism. The threat primarily manifests itself in the activities of environmental actors who believe they are responsible to rules and regulations that trump activities, including those oriented toward freedom, equality, and social justice.

Chantal Mouffe states an undeniable fact, "We can never be completely satisfied that we have made good choices since a decision in favor of some alternative is always at the detriment of another one" (2000, 135). In general, the absence of satisfaction is not considered by theorists of green citizenship. Green citizenship needs to render politics central to its development if it is to take seriously its participation within democratic communities; by doing so, green citizenship can contribute to the expansion of a sphere of activity that is never complete. Furthermore, once situated in democratic discussions, green citizenship can allow environmentalism to take its place alongside the plurality of movements resisting the gradual elimination of all spaces of wonder, autonomy, imagination, and utopian vision. A central place for politics means green citizenship may not encompass all the needs and desires of environmentalists, but it also means that green citizenship can become something distinct within both green and democratic public spheres.[5]

The absence of a strong-radical democratic presence within green citizenship discourse has meant that environmentalists who claim to be green citizens have kept to the moral high ground and thus not situated their ideas and goals adequately along a chain of equivalence. As a result, when it comes to green citizenship notions such as sustainability, the politics of ecology, and the rather mixed history of environmentalism with its strands of racism, sexism, and at times explicit authoritarianism, have eluded critical interrogation.[6] Once again, the way to address this tendency within environmentalism is to perform as committed democratic citizens not subject to specific disciplinary goals or purposes.

In the remainder of this chapter I critically interrogate and point to the democratic deficiencies of what I see as the main approaches to green citizenship. I also, however, continue to defend the idea of keeping the discussion around politicizing nature active both as a disruptive participant within citizenship theory and as a disturbance within green public spheres.

While there are many ways of dividing green citizenship discourse, I have chosen to divide the remaining section into three parts. I begin with stewardship as I believe the stewardship ethic, as a part of green citizenship, is the most dangerous and threatening in its colonization of the political and its use of present political institutions and authorities to achieve its predetermined and fixed needs. While it is not necessarily authoritative, stewardship is normative, requires expertise, has a heavy reliance on the state, and is rarely if at all interested in democratic ideals of freedom and equality. Following the critical interrogation of stewardship as a partner to citizenship, I turn to the care ethic. At present, nearly every defender of green citizenship believes a personalized ethic of care is an essential component of green citizenship. While not denying the potential value of an ethic of care as a stimulus to environmental action, I argue that a commitment to citizenship, if it is radically democratic, embodies its own broad ethic of respect that extends or replicates a compassion-free ethic of care that includes a need to respect the value of others' opinions, yet frees the actor from the need to know or like the other who is being respected. In the final section, I look to ecodeliberative theorists as they represent a wing of environmental politics that at least attempts to take the turn to politics seriously.

The Stewardship Ethic

Supporters of the stewardship ethic offer a prime example of the difficulty in trying to reconcile a democratic ethic with an ecological ethic. Here, highly virtuous and ecologically knowledgeable citizens and states are to watch over and guide other humans and nonhumans in order to resist the ensuing tragedy (informed by scientific truths). Theorists like Christoff, Barry, and Newby look to the citizen and the state as green partners who may, if adequately influenced, resist the present drive toward ecological Armageddon. Their shared belief is that we all face a state of crisis that threatens both humanity and the rest of nature and "that such problems and threats require urgent resolution" (Christoff 1996, 158).

As the informal activity of democratic citizens cannot offer ways to achieve such resolution, they have no choice but to turn to more authoritative means, including support for a "reconfiguration of the state to provide widespread guarantees of environmental rights" (Christoff 1996, 157), or a form of representative democracy[7] combined with an emphasis on management, decision making, and end-focused deliberation (Barry 1999).[8] While they are aware of the cosmopolitan and nonreciprocal nature of Dobson's ecological citizenship, supporters of the stewardship ethic nevertheless focus on more formal state-citizen relations as, according to them, the state remains an important actor within international bodies, and has the authoritative means to ensure new responsibilities can be met. The state and citizens of the state become stewards and watchdogs with knowledge and power that can guide, or are imposed on, everyone's activities. This kind of stewardship ethic has particularly troubling consequences when situated in citizenship discourse. In essence, it finds in the relationships between the state and citizens, and citizenship and responsibility, an opportunity to use already existing institutions and responsibilities as ways of ensuring stewardship becomes a part of collective and individual political activity.

For example, Barry (1996, 125), who applauds the practicality of the stewardship ethic, explains that "in comparison with anarchistic versions of green politics," a stewardship ethic is "compatible with and indeed require[s] a commitment to state or state-like institutions," due to the fact that "the state is a necessary (though not sufficient) condition for the elaboration of discourse of sustainability in the public sphere of modern liberal democracies." In a reciprocal relation between the state and the citizen, the latter is committed to influencing the state in much the same way as the state is committed to creating responsible citizens. If successful, this relationship will ensure human society takes on its role as steward rather than oppressor of the earth.

Christoff (1996) similarly recognizes the value of the state as the primary actor in the international community and likewise calls for ecological citizens to help in the "restructuring of the state" (1996, 151) in order to create a morally responsible citizenry informed by an "environmental constituency [that] includes all those with an identifiable vital interest in the outcome" (1996, 156).[9] Specifically, Christoff (1996, 163) wishes to institute an ecologically guided democracy with a "hierarchy of values" that replaces "narrow anthropocentric values"—like the rights of individual humans, classes, or nations—with "universal ecological values or principles." States and citizens are to be guided by these principles while respectfully performing their duties as responsible international actors and self-disciplined ecological selves.

So an *ecological contract*[10] between the state and the citizen, such as that proposed by ecological stewards, requires a state willing to use its authority to ensure green outcomes and a virtuous citizenry willing to accept the role of ecological steward. Both are to be capable of stewarding nature, which in turn must be seen as in need of human assistance.[11] Absent from the equation are the explicitly political questions concerning how humanity can ever know how to steward nonhumans, who and what is to inform the stewards, how the informing knowledge is constructed, and why and how stewardship is an act of citizenship. Furthermore, even in the unlikely case that the state does become green and citizens do become virtuous and responsible,[12] there remains no discussion of the place of freedom, politics, or active participation in the resulting society. Moreover, humans are capable of and should be asked to do much more than adapt or steward. There is no reason to think that individual humans should not be given the opportunity to flourish in much the same way as the rest of nature. The goal, or democratic calling of green citizenship, should include promoting both the flourishing of human potential and the flourishing of nonhuman communities. There is no need to view these two goals as oppositional; there is, however, good reason to view them as antagonistic.

According to Barry, "being a good citizen from a green [stewardship] point of view consists not merely in considering the interests of non-citizens in making environmental choices, but also in acting in a manner which promotes ecological stewardship" (1999, 235). Thus, what being a good steward, trustee, or citizen is, along with what a well-stewarded society might entail, are issues that should be up for debate and will lead to antagonism between participants. The debate, however, is dependent on political discourse that does not prioritize or defend stewardship. Debate on what good stewardship could be is political and relies on citizens who ask critical questions grounded in the ethics of democracy. Therefore, if supporters of the stewardship ethic are going to situate discussions around stewardship in the public sphere they need to recognize the primacy of democracy and the accompanying ethic of freedom and equality. Stewardship may not be essentially antidemocratic but, it certainly does not, in its current manifestation within green citizenship discourse, contribute to the forces of freedom or the expansion of the public sphere.

From this brief overview we can see that the stewardship ethic manages to be anthropocentric and antihumanist at the same time. While it demands disciplined human stewards of the earth, it also supports a highly behavioral "self-limiting culture of moderation and responsibility, producing actors whose environmental awareness would morally and materially confine their actions to those producing ecologically sustainable outcomes" (Christoff 1996, 166). "End-focused" projects typically require that people give up agency to facilitate better and more immediate management. This loss of agency often leads to greater efficiency, better control, superior incarceration, and so on, but if agency is lost so too is the promise of politics as a sphere of freedom, action, and, perhaps most importantly, natality for it is within the capacity to bring something new into the world that environmentalism's political promise has always lied. Consequently, the sacrifice is too much to ask.

What the stewardship position best illustrates is the magnitude of the position with which one enters the political sphere. If, as with the stewardship ethic, one enters with a fixed prepolitical belief, there is little chance for democratic citizenship to remain a significant component of the ethic for it would be irrational to dabble with freedom when faced with the needs of survival.[13] The ecological position is dominant and the political citizen is seen as an empty vessel to be permanently filled by a disciplined, virtuous, and ecologically informed ethic. But unfortunately, and more to the point, the highly disciplinary ecological position is touted as an appropriate foundation for citizenship. The empty space of political power created and performed by citizens and opened up by the loss of the markers of certainty, is to be permanently filled by a new highly disciplinary ethic that trumps political desires not because it is more just, but because it is more necessary.

Once the partnership between ecology and democracy is scrutinized, it becomes clear that there is a necessary and essential tension between active democratic citizenship and the sort of adaptive disciplinary management required for successful stewardship. Fortunately, this irreconcilable tension has been considered by democratically inspired green theorists who along with examining how ecological issues could impact political discourse also focus on the consequences of grounding ecological knowledge in radically democratic political spaces.[14] I will look at a particularly potent group of these theorists later in the chapter; suffice it to say here that at best the stewardship ethic may be a good way to *work* with nature outside the sphere of politics. Compatibility between stewardship and democratic citizenship, however, is a fallacy with antidemocratic threats far too great to ignore.

If citizenship is to remain in democratic communities, no matter how important each disturbance is, the true green-democratic challenge must always bring one back to the question of how to make obligations to "animals, trees, mountains, oceans, and other members of the biotic community" (Smith 1998, 99), a duty of care owed to "the planet in terms of minimizing resource consumption and pollution"

(Steward 1991, 74–75), and a commitment to "act with care and compassion toward distant strangers, human and nonhuman, in space and time" (Dobson 1999, 8) fit with a commitment to democracy. Such commitment, as we have seen, means that "green outcomes cannot be guaranteed" (Arias-Maldonado 2001, 12). Environmentalists must *stop and think* about the possible consequences of their position. It cannot be forgotten that the task of the citizen is not to be responsible to the environmentalist,[15] regardless of how needy the environment may appear.

The Caring Ethic

The most common ethical position held by those promoting green citizenship is the ethic of care. What distinguishes these theorists is that most have *stopped to think* about their arguments. What makes them the same as other theorists of green citizenship is that they too tend to bypass the political sphere and focus primarily on the private and administrative spheres.[16] That is, while they move beyond the macroethics of responsibility and stewardship, they stop short of moving beyond the *need* for citizens to adapt to ecological imperatives rather than create their political identity.[17] Nevertheless, it is what distinguishes their position that makes the care ethic most relevant *and* most dangerous.

Lip service is paid to an ethics of care by most theorists who support a responsibility or remoralizing focus of green citizenship. Among these thinkers is Hartley Dean (2001, 501), who argues that an ethic of care "provides the crucial link between an abstract position of co-responsibility and the substantive practice by which we continually negotiate our rights and duties." Alternatively, Andrew Dobson (1999, 25) suggests ecological citizenship is about the "virtues of care and compassion, practiced in both the private and public spheres," and Fred Steward (1991, 74–75) believes we "owe a duty of care to the planet in terms of minimizing resource consumption and pollution." The ethics of care is best represented, however, by Deane Curtin.

Curtin is refreshingly honest about his allegiance to a critical ecocommunitarian ethic when he argues that "the best *guarantee* we have of preserving the wilderness of nature is through cultivating an informed and humble [local] citizenry that is genuinely committed to preservation" (Curtin 1999, 190, emphasis added). The way to create such a citizenry, Curtin believes, is to have a sense of home that allows one to put down roots, which "define who we are, [and] where we belong; it also defines who we are not, places where we are strangers" (1999, 174). Curtin recognizes that such an ideal will be difficult for his North American audience; according to him, we are "perhaps the lone people that has never succeeded in becoming native to any place" (1999, 173); indeed, "we in the west are homeless" (1999, 170).

A warning flag should be raised any time a guarantee is sought within a democratic ethic. And one should ask why being placeless is necessarily a failure.[18] Nevertheless, such concerns should not deter the reader from further considering Curtin's approach to ecological citizenship as it is embedded in a useful and necessary critique of the hegemony of Western domination in relation to both the nonhuman other and non-Western cultures.

For Curtin, we in the West must embrace or initiate an ethical transformation in order to become invested in our place while also respecting the place of others. Such an ideal of "principled engagement in community" (1999, 186) is to be informed by "a pluralist ethic that begins with the *authority* of local communities to define their local values and participate in their transformation over time" (Curtin 1999, 141–42, emphasis added). Curtin's care ethic is concerned with an authentic environmental ethic and is founded on the belief that "care for and understanding of nature must come to function as an internal good, constitutive of what it means to function as a citizen" (1999, 179). He suggests that placeless North Americans should turn to a mix of Edward Abbey, Thomas Jefferson, and Wendell Barry—all three of whom may not only have been American celebrants of place but who also had particular deficits in relation to democratic ideals—as their authentic guides.

How each community—itself diverse and plural—is to achieve such authority is left to the reader's imagination, which is certainly acceptable as articulating any democratic form or procedure would surely be the responsibility of the community itself. But, if one is committed to democratic ethics, one is also committed to the belief that authority is not what emerges from democratic engagement. What emerges may be a better understanding of a variety of issues and interests, which in turn would lead to the recognition that authority is not a gift democracy is capable of delivering.

Along with a desire for frictionless (harmonious) relations,[19] a major problem with the care ethic is that it shares both the slipperiness and the positive connotations of democracy. As such, a reasoned critique of its merits is very difficult. If one does not situate the ethics of care in the political sphere where one can *stop and think* about the potentials and limitations of the ethic, it can easily become a dominant position difficult to resist, as resistance would seem *un*caring. The fallacy is, once again, a tension-free relationship where humans and nonhumans dwell together on the land, and an apparently needy nature is cared for by virtuous human actors. While Curtin does attempt to politicize his ethic of care, he fails to achieve the politicization as he enters the communicative moment with too strong a commitment to care and thus lacks the desire or ability to actively listen to other worthy positions.

The sort of respect necessary for active listening may be impossible beside a strong commitment to an ethics of care. Indeed, respecting others may mean recognizing our inability to know nature well enough to be sure that what we perceive as care is not actually another form of domination or exploitation obscured by an arrogant and self-satisfying belief in a needy nature that can be assisted by our altruistic desire to live in harmony with nature. As Curtin spends much of his book, *Chinnagrounder's Challenge: The Challenge of Ecological Citizenship*, speaking of the well intentioned yet colonizing exports of Western ethics into "third world countries," it is surprising that he does not acknowledge this potential outcome or arrogance. As he rightly explains,

> It is often the well-meaning assumption that "we're all the same," that we must share some common culture, or the same basic human nature, if we are to communicate at all which corrupts the possibility of genuine communication. The very expectation that we will find deep reservoirs of commonality may cause us to invent what we want to find, and thereby ignore serious points of divergence. It is only when we begin from the possibility that others are unimaginably different, that we are open to the surprise of genuine understanding. (Curtin 1999, 162)

Curtin (1999, 171) also describes a unique encounter a friend of his had with Adivasi people in southeast India that gives what I believe to be the most well-articulated argument against Curtin's own ethic of care. The story was based on a question Curtin's friend asked concerning a boulder that blocked easy access to the Malayali villages.[20] The friend asked, why not simply push the rock down the valley? The response he received from the villager who was "clearly taken aback" was "just as all people have their proper place in the world, so the boulder has its proper place, and it would not be right to move it because it is unsafe for human beings." There are two key messages here. The first is that there is no mention of caring or partnership by the villager. Rather, there is recognition of otherness and the significance of minimizing our involvement whether it is for so-called altruistic or selfish reasons.[21] The second message is that, had the *caring* Westerner not had the communicative moment he may well have "helped" the villagers by removing the rock. Surely, the same well-intentioned care of nonhuman community members could have equally negative consequences.

Crossing the species boundary makes communication even more difficult, and active listening even more relevant,[22] but however actively one listens to the nonhumans it will always be done through numerous translation filters. Each of the filters distorts the message. The inability to know nature is what makes nature so wonderful, so wild, and so vital as an *other*. Citizens committed to democracy ought to be more concerned with allowing that part of nature unknown, indeed not knowable, to flourish.[23] The democratically informed green citizen need not be bracketed out of the political community. However, s/he would be expected to participate as a democratic citizen, and thus humble amateur, whose contribution would be more toward expanding agonistic democracy than protecting nature (it is only by emphasizing the latter that the two can be thought of as mutually exclusive).

Gazing upon and interrogating an ethic of care with a critical democratic eye does not render an ethics of care useless to a democratic ethic. On the contrary, there is no reason why a personal ethics of care could not be compatible with a public democratic ethic; indeed, it may be the best potential for cooperation between the public and private spheres. Like Dobson's argument in support of the potential of the household as a place where the virtues of ecological citizenship are created, an ethic of care is an explicitly personal ethic,[24] but one that *can* exist alongside many other ethics within a radically democratic society.[25]

For citizens, as opposed to caregivers, allegiance is always first and foremost to democratic ethics; this relation is not necessarily about caring. The democratic citizen is required to respect others whether or not s/he cares for them, and the public realm "simply cannot afford to give primary concern to individual lives and interests" connected with individual care (Arendt 1958, 156). In fact, how much one cares for the other must be bracketed out of the political sphere, as political engagement requires everyone to be given equal respect. Furthermore, to care, or compassionately assist, seems connected to a need to know the other, and the only way individuals can know the other is to translate the other into a discourse where they may uncover *what* the other is but never *who* the other is. Stated slightly differently, while we may know *what* a nonhuman animal is by increased place sensibilities (or more likely uncovering its DNA, domesticating it, and/or sticking it in a zoo or circus for our *needs* or amusements), we can never know *who* a nonhuman animal is.[26] Recognizing this fallibility is not to say we simply ignore the nonhuman. Rather, it is to say we respect the nonhuman at least partially as an unknowable and wild other. It is to say we allow the nonhuman other the opportunity to flourish by expanding democratic deference to include our interactions with the nonhuman without forcing it to gain agency only through human classification.[27]

It is in relation to this need for communicative action that the next section on "ecodeliberative ethics" is most relevant. But before moving on to ecodeliberative ethic's attempted interspecies communicative approach to representations of nature, it is worth pausing for a moment to examine the only other approach to ecological citizenship that attempts to free itself from the tradition of ecological thought. That approach is given by Herbert Reid and Betsy Taylor (2000) who, unlike Curtin (who turns to American theorists of place to support his position), use empirical research to uncover the problem with North American placelessness, arrogance, and homogenization.

By grounding their definition of ecological citizenship in surveys of real people within an actual community, they offer a specific example of interaction between humans and nonhumans that does not require the citizen to be "symbolically produced as an abstracted, universalized ideal out of the messy materiality of particularity and difference" (Reid and Taylor 2000, 453). In fact, it is the uncovering of particularity and the implicit emphasis on the need to retain particularity that makes their contribution to the discourse so germane.

Briefly, the empirical research they undertook found that among Appalachian mountain dwellers, there was a bioregional imaginary that led them to consider walking and storytelling as types of environmental activism and citizenship. Not surprisingly, those who self-identified as "environmentalists" but did not dwell and live on the land made no mention of these activities as types of environmental activism.[28] In fact, nonnative dwellers "tended to foreground goal-driven activities like recycling, political lobbying, and green shopping" (Reid and Taylor 2000, 460) as the way to enact their environmental consciousness. For Reid and Taylor (2000, 440) "to refer to 'ecological citizenship' is to speak of citizens embodying the particular places of ecological experience with common concerns (and concerns grounded in the commons) potentially leading to expansive spatiotemporal horizons of responsible action." Their concern is with finding a concept of citizenship that recognizes that "we are dwellers on the land" in much the same way as Aldo Leopold wanted humans to become plain citizens of the earth. Similar to Curtin, they find a strong relationship to place as essential to ecological citizenship. They also share his critique of mainstream Western culture and environmentalism.

There are, however, two related and particularly noteworthy findings that come out of their research, and distinguish their insights from Curtin's.[29] The first is the peculiarity of their discovery that walking is considered a type of ecological citizenship among Appalachian mountain dwellers.[30] Recognizing walking as an informing or potential act of citizenship already disturbs modern notions of what counts as politics in a placeless world where Western man's place is every place (Curtin 1999, 175). It also brings to mind Thoreau's wonderful essay on walking where he makes his most significant, yet largely forgotten claim that, "in Wildness is the preservation of the world."[31]

The reason Appalachian mountain people could argue that walking was a significant part of their environmental activism was because it was undertaken for many reasons including to "see if anything has changed" and to "see if anything was new" (Reid and Taylor 2000, 459), and then to share stories as "walking is as much a labor of story building as it is of productive activity" (Reid and Taylor 2000, 461). The connection between walking and storytelling only occurred, indeed can only occur, among those who had deep connections to the Appalachian mountains.[32]

It is this connection between walking and storytelling and the peculiarity of the walking (and joyful) experience or stimulus for engaged discourse that makes Reid and Taylor's study particularly important to thinking about green citizenship. It is the storytelling that is the primary act of citizenship; it is the walking that informs it. This can be read as an indirect argument in support of a revitalized public sphere where private stories can be publicized in open space and retold in light of opinionated others with their own stories to add to and/or refute the stories of others.

Storytelling resists homogenization as it gives places "depth and resonance because they are saturated with the intersection of story upon story" (Reid and Taylor 2000, 461). It also allows the particularity of a place to be shared and added to with each additional story. Stories are not grounded in truth or bound up in a desire or need to achieve a greater goal; they are intrinsically valuable, opinionated, and purposely or inevitably situational.[33] As dwellers live and work the land, walking is both productive and intrinsically valuable: "Walking the hills inextricably mixes material extraction...with a profound savoring of bio-aesthetic engagement with the wild" (Reid and Taylor 2000, 461). What is most important here is that they experience and en*joy* the land differently. There is no need to argue that their way of life is better. There is, however, a need to respect and protect their otherness in much the same way as we must respect the otherness of nature.

Allowing walking and storytelling to be a part of one's environmental activism offers an approach to citizenship at least partially akin to the joyful way Hannah Arendt sees politics. There is joy in walking and joy in sharing stories about the discoveries experienced while walking. It is not that walking *is* political but rather that walking *can be* politically relevant and can inform one's actions as a citizen. Here, green citizenship is informed by activities that do not fit neatly into any category. The act of walking can end up being productive or not. It can lead to discoveries, or not. It can inform political action, or not. It is both public and private and neither public nor private.

What is certain is that typical goal-oriented environmentalism rarely offers such *wonder*ful potential. This is why the most important lesson from this study is the need for continued expansion of what *can be* political. As the only way the discoveries are made is by allowing nondirectional wandering, both figuratively and literally, to be a part of one's politics. Theorists of green citizenship can learn that they must be careful not to colonize the political sphere with goal-oriented desires. This finding drastically disturbs any typically Western notion of citizenship and requires readers to *stop, think, and ask more questions* in order to understand how one could view walking as an act of green citizenship. As the simple act of stopping and thinking is a necessary component of citizenship, we can say that Reid and Taylor have already done a service to citizenship discourse by bringing this idea out of obscurity.

Additionally, their research illustrates that only the dwellers of the land, often thought of as antienvironmental by those who view nature as a place to visit and protect as other, spoke of walking as an environmental activity.[34] This suggests the need for plurality and discursive democracy rather than the reverse need for procedures that will legitimize prepolitical ecological knowledge. That is, the results from their research can be interpreted as support for democracy, communicative action, active listening, a focus on understanding, and the need to recognize the inevitable limitation of every construction of green citizenship. Plurality and political action are essential partners. The loss of one will lead to the loss of the other.

The point I am making here is that while uncovering walking as an act of ecological citizenship was extremely important—certainly it disturbs common notions of citizenship and has much to teach theorists of green citizenship—it is not merely the peculiarity of walking that is significant. It is also relevant that before the unique study, and by extension without the unique study, the notion of walking as an act of citizenship was simply unthought of. As already noted, none of the nonnative Appalachian dwellers mentioned anything like walking in their approaches to environmental activism. It is not that walking is necessarily a *better* way to become an active green citizen. It is simply that, it is *different*, and adds to the political argument in defense of actively listening to and searching out a plurality of stimuli to ecological knowledge. Furthermore, their study shows that there remain *traces of joyful citizenship* among those not as explicitly, or at least differently subjected to capitalist and statist relations.

The research uncovers the obvious yet often-obscured fact that Western notions of ecological citizenship are not "good enough." In fact, like all other normalized identities, the Western citizen (even in its ecological manifestation) is a particular and situationally specific political agent that has attained nearly universal status. What Curtin, and Reid and Taylor's research offers is an implicit defense of the position that when we do represent nature it should be within informal political communities oriented toward understanding how nonhumans interests, which we can never truly know, can be situated within a multiplicity of interests and opinions, including those which are not necessarily ecological.

On this note, we can turn to the last group of thinkers who find promise in the rise in deliberative democracy and discursively legitimized knowledge.

Ecodeliberation: The Deliberative Turn

Dean (2001, 500) in his ecosocialist contribution to green citizenship discourse supports a position founded upon two ethical premises. The first relates to the distribution of scarce resources, and the second to the provision of care. As we have already discussed the ethic of care, I turn here to the first premise.

Dean's focus on distribution is supported by a macroethic that "combines social justice with ecological sustainability" that can bind humanity in such a way as to enable it to collectively address the "environmental predicament." He approaches this predicament by way of a "planetary principle of co-responsibility" founded on three conditions. First, the ethic must be rational (discursive) and rise above tradition. Second, a global communications community can and should be used as a means for creating the potential for a Habermasian ideal speech situation (ISS), which allows for the *unforced force of the better argument* to emerge from uncoerced and transparent communication between humans and nonhumans. Third, resonating with Ulrich Beck (1992), and most at odds with the democratic form of citizenship I have been defending, the principle must take scientific and ethical *claims to truth* equally seriously so different epistemologies can be negotiated on equal grounds.

These three ethical premises place Dean's arguments within a rich history of eco-communicative discourse quickly becoming as dominant within present ecological political theory as deliberative democracy is within democratic theory. There is a substantial list of green political theorists who refer to this democratic approach including the already discussed ideas of John Barry and Peter Christoff. But when the focus is on interspecies communication as a means to expand democratic consideration, the most interesting and well-developed theories of ecodeliberation come from those who never explicitly call for green citizenship. The defense of deliberative democracy is found primarily within the arguments of a number of theorists deeply involved in the journal *Environmental Politics*, all of whom recognize that "the structure of liberal democracy itself is ultimately incapable of responding effectively to ecological problems" (Dryzek 2000, 143). In this section, I look primarily at three particularly significant voices who guide the discourse, John Dryzek, Val Plumwood, and Robyn Eckersley. Dryzek and Plumwood have each written books on the topic and Eckersley has of late changed her position from a strong critique of communicative ethics (1990)[35] to a position of critical support (1999).

What makes these theorists slightly different from the previous contributors to the discourse is that they begin their ecodemocratic theory as much from a democratic ethic as an ecological one. Why they turn to the radical democratic theory of Habermasian discourse ethics will become clearer shortly, suffice it to say here that deliberative democracy's end-focused and consensus-oriented discourse requires far fewer limitations on epistemological truths (including ecological ones) than do the agonistic theories of Arendt and Mouffe.

By seeing the discourse ethic as "more likely to lead to the protection of common or generalizable interests" (Eckersley 1999, 24), the truth focus of the representations of nature is left uncriticized from a political perspective. While *particular* truths are challenged and interrogated, the commitment to *truthfulness* is not. Furthermore, the ecocentric desire for tension-free partnership between humans and nonhumans remains a central feature of the eco-communicative ethic, all but eliminating the possibility of embracing the tension and recognizing the limits of the human ability to know nature.[36]

Nevertheless, deliberation takes politics out of the expert-driven sphere of administrative decision making and places it squarely within the realm of amateur communication where what is most relevant is reaching understanding, achieving agreeable consensus, and guiding legitimate legislative action.[37] Thus, there is solid reasoning behind the eco-communitarian's turn to discourse ethics as there are conditions for participation in this sphere that offer extended possibilities for the inclusion of representations of nature. These conditions, when partnered with Habermas's deliberative democracy, include a desire for consensus; openness to discursifying traditions previously considered untouchable; acceptance of another's ideas rather than status; and a desire for inclusivity. It is through the realization of these conditions for an ISS that the unforced force of the better argument can prevail and legitimate decisions can be made. According to Dryzek (2000, 147) achieving anything like an interspecies ISS would include "detaching democracy from liberal anthropocentrism, while retaining an emphasis on deliberation and communication." It would also surely include a rather remarkable ability to have uncoercive, intersubjective communication with nonhuman animals.

The environmental interest in this procedural approach lies in the potential, and arguably need, for including representations of nature in discursive communities. By attempting to cross species boundaries, each theorist also attempts to, as Dryzek puts it, save communicative ethics from Habermas. The need to disarticulate communicative ethics from Habermas is due to Habermas's position that interaction with nature can only take place instrumentally (manipulation and control) as we can never know the interests of nature. The way they attempt to save nature from the instrumental gaze is to insist that nature already is social (Vogel in Eckersley 1999, 38), which means rejecting the nature-society divide and accepting that the social construction of nature itself needs to be subject to communicative ethics. Once again, this is a fair and acceptable argument even from a radically democratic and agonistic perspective. What is problematic is where the ecodeliberative theorists take their reasonable argument.

They argue that the inclusion of the non-humans can be achieved through extending the deliberative requirement for inclusion in order to accept not only those with linguistic capacity but also those with agency, intentionality, or the capacity for self-directedness (Eckersley 1999, 25; Plumwood 2002, 175).[38] Such arguments become possible once the politics of *ideas over presence* is accepted so that nonhumans can be received as "imagined partners in conversation" (Eckersley 1999, 27). Once a deliberative community is based on attempts to achieve an ideal speech act, all participants must act *as if* it were actualized; in this context, representations of nature can attain equal status.[39] Here, those who wish to represent nature extend the political community in order to include the interests of the nonhumans among those impacted by decisions. The ecodeliberative theorists wish to use their arguments to create a speech situation where the interests of nature are always considered.

By situating environmental claims in deliberative or communicative communities, they are rendered equal. This sort of attempted inclusion is certainly an improvement over present liberal democratic institutions. However, even in the communicative model, to attain space in this community the nonhuman must be represented by humans. As this is considered a problem to be dealt with as opposed to an inevitability to be explored, the main concern becomes how to decide who speaks for nature and/or how to listen to its messages. The desire becomes how to improve the accuracy and legitimacy of the representation so as to help legitimize first, the inclusion of the nonhumans in deliberative bodies and second, to help those bodies achieve an interspecies ISS.

Earlier it was shown that Curtin and Reid and Taylor believe that the best way to hear and understand the interest of the nonhuman is to become native to place. While highly contentious, this is a position shared here. Further remoteness, they convincingly argue, is not only a problem concerning interactions with nature, but it is also a problem for those detached from a diversity of political discourse, and the consequences of the decisions they make. Their principles lead them to "criticize institutions that try to subordinate nature on a large scale, and those that are remote and so incapable of hearing news from below" (Dryzek 2000, 154).[40] Yet, they also recognize that remoteness from political ideas and advances is just as real a threat to "outside" communities as remoteness to nature is to those living in cities (Plumwood 2002).[41] In both cases, the problem relates to the inability or refusal to listen.[42]

On this issue Dryzek (2000) argues the concern is not so much with silencing the other as "nature 'speaks' or does not 'speak' irrespective of any attempted human suppression of that ability." What we must do is learn how to understand what we are being told. For Dryzek, appropriate listening must be an eminently rational affair so we can "listen to signals emanating from the natural world with the same sort of respect we accord communication emanating from human subjects, and as requiring equally careful interpretation" (149).

For Plumwood (2002, 169) learning to listen would include developing

> narrative and communicative ethics and responses to the other, developing care and guardianship ethics, developing alternative conceptions of human virtue that include care for the non-human world, and developing dialogical ethical ontologies that make available richer and less reductive ways to individuate, configure and describe the world that "make the most" of the non-human other.

She also supposes that if we are to have interspecies communication "we humans must learn to communicate with other species on their terms in their own languages or in common terms, if there are any" (Plumwood 2002, 189). It is presumed, once again, that the way to do this is to live in a particular place and interact in interspecies communities.

Yet surely, what must accompany our ability to listen to nature is the recognition that we never actually hear it correctly. Even when we do hear it, the voices tend to be translated into needs not desires. As already argued, it is reasonable to imagine that in certain cases we can begin to know *what* a species of animal needs to survive, or even what a particular animal dislikes. We cannot know the animal's desires or *who* the animal is.

There is grave danger in overemphasizing the need to listen to nature without also insisting on public space for those messages to be discussed as particular interpretations of particular messages. Habermas (Whitworth 2001, 26) himself understood that "the more pluralist a society becomes, the smaller the number of morally resolvable questions"—which is where he should have ended. His problem, one shared by ecological democrats who follow his lead, is that he continued by arguing that this makes it all the more crucial "to resolve them." Namely, communicative space oriented toward reaching understanding remains tied to deliberative procedures oriented toward legislating legitimate decisions. This linear movement brackets out all those stories that do not conform to the interests of the majority. Interpretations of nature are always limited and no matter how well one listens, what each person hears will be different. In addition, there is nothing to say that the messages heard by those closer to nature are necessarily more accurate. If nature speaks to us it can also dupe us and we certainly cannot be sure that the message we believe we are receiving is the actual message being sent.

One should certainly listen to nature, but we should not try to resolve moral questions or reconcile the inevitable tension between humans and unknowable nonhumans. Furthermore, if we really want to increase the truthfulness of our representations of nature we must listen to other's stories about nature as much as we should try to improve our own ability to hear the messages.[43] Truthfulness may well increase relative to the number of perspectives one is capable of understanding but never does one achieve *the truth*. And when it comes to representations of nature one must be very careful not to think the capabilities of listening are enough to *uncover nature's secrets*. Resolutions, even without interspecies communication, may be necessary, but must only, in an agonistic democracy, ever be accepted as temporary. Resolutions must be open to political scrutiny; meaning, once again, a primary concern of those interested in eliminating oppressive and domineering relations with nature is in the expansion and permanence of the political sphere.

As deliberative democracy is the present darling of political theory, it is no surprise that environmentalists are struggling for access to this discourse. But as I have argued, the democratic potential of environmentalism lies not in participation within limited decision making bodies, but rather in disturbing the legitimacy of these bodies in order to stimulate discussion around alternatives. What environmentalists should do is acknowledge and accept that their desire to put an end to the oppressive and domineering relationship between human and nonhuman is, or could be, an extension of a broader democratic ethic. I am not speaking of an extended self, but rather a crossing and challenging of the boundaries of the sphere of concern. It is not however a crossing of the participatory boundary. Nor is it a crossing for compassionate reasons. It is for the democratic reasons of rightness or justice.

Bringing this section to a close, we can say that the promise of the deliberative turn lies not in improving our ability to listen to the nonhumans, but in disrupting the species boundary and expanding what are considered legitimate stories to be told. While this procedural approach to democracy may, it need not lead to more rational and legitimate decision making bodies. There is certainly just as much potential of this expansion leading to the paradoxical conclusion that legitimate decisions are simply unattainable. Accepting this is not accepting failure as much as it is recognizing the dire need to continue expanding and protecting the specificity of the political sphere as a space oriented toward reaching *unattainable* understanding.

Humbling the Green Citizen

So why consider green citizenship at all? What could possibly come from such a flawed starting point? Isn't the whole problem the term "green"? All these questions are fair and should enter the political discussion around the promise and limit of green citizenship, perhaps as part of a new green public sphere. But along with such pessimistic questions we should also ask, what if the strong ecological starting point and disciplinary focus do not remain central to green citizenship? What if the articulation between green and democratic political theory is not limited to current views? What if the skew in the relationship could be moved in the direction of democratic citizenship rather than green ecological desires? In this chapter, I have shown that if citizenship continues down any one of the current paths it is traveling, there is little obvious liberatory promise. If, however, as Mouffe has taught us, historical articulations are not fixed, the rhetoric of green citizenship need not be left in its present state. There is disruptive promise in green citizenship, it just must be realized that a large part of the disruption is political.

As Bookchin (1995, 47–48) has explained, "humans are vastly different from other animals in that they do more than merely *adapt* to the world around them; they *innovate* and create a new world, not only to discover their own powers as human beings but to make the world around them more suitable for their own development, both as individuals and as a species" (emphasis original). It is as a democratic citizen that this human potential is unleashed. I have argued that those promoting green citizenship focus too much on how to make citizens act according to a grander imperative. This focus has obscured the latent liberatory opportunity opened up by situating representations of nature in political discourse. Rather than participating as an equal in a field of discourse all-too-often, the green position is considered above all other concerns. When such a disciplined allegiance guides the discourse, the potential to innovate is lost and politics is all but eliminated. At no time are citizens given the opportunity to strive toward their own democratic potential.

More recent work on ecological citizenship has created a new terrain; one that feral citizenship might contribute to. These theorists are part of a small group of what might be best referred to as friendly critics who wish to engage in the green public sphere and I believe keep citizenship central to the discussions that constitute the green public sphere. Following in the footsteps of Barry (2006) with his sustainability citizenship that challenged proponents of environmental citizenship to expand the prarameters of the field of justice, and theorists like MacGregor (2001; 2004; 2006) and Agyerman and Evans (2006), whom have brought social justice issues to the conversation around environmental citizenship in order to point to the problems with the often myopic nature of environmentalists, these friendly critics are starting to draw attention to the baggage that continues to accompany the environmentalist turn to citizenship. Without always emphasizing it, Latta (2007), Hayward (2006a), Gabrielson (2008), and Gabrielson and Parady (2010) have all pointed to the lack of politics and self-reflexivity among those promoting the partnership between environmentalism and political agency. The partnership is promising, but the liberatory promise requires taking the turn to politics seriously and it requires allowing that turn to push and challenge the boundaries of the political at the same time as accepting some of these boundaries if only temporarily.

It has been shown that by adding green, ecological, or environmental qualifiers to citizenship the citizen is often reduced to an agent of the environmentalist. The fallacy of equal partnership *is* simply a fallacy. This is why I have argued that if theorists of green citizenship are to take seriously their commitment to democracy they must give up many of the pillars that previously supplied legitimacy for participation in authoritative bodies. There is no doubt that it is dangerous. There is no doubt that such a commitment to democracy will, at times, result in "environmentally unfriendly acts" but there is also no doubt that without such commitment green citizenship will never achieve its democratic potential. Furthermore, there is ample reason to believe the more democratic a society, the more environmental it will be.

It is only a radically democratic society that can welcome the tensions between green and free citizenship and accept them as inevitable and potentially liberatory. By grounding tensions between humans and nonhumans in agonistic political discourse, we can eliminate the desire to solve the tension through stewardship, care, compassion, or some other false partnership while concurrently ensuring the tension does not digress into violent or oppressive relations.

We have seen that the main problem with representation lies in its inability to ever be accurate. The *representer*, by definition, is to become more than or other than herself while representing the interest of another. The job is highly disciplinary and lacks the joy, creativity, and spontaneity of the radically democratic citizen. The purpose, and thus the end point are more relevant than the political act. So once again, if environmentalists are serious about democratic citizenship in their desires to politicize nature, they must not try to speak for nature or attempt to become nature rendered self conscious. Such an attempt, whether from scientists, activists,[44] or ethicists will always result in failure and will never achieve the required legitimacy for participation in more managerial politics.

Finally, if agonistic pluralism is indeed necessary for a healthy democracy, as I have argued it is, and green citizenship is to become a part of rather than a replacement to radical democratic projects, as I think it should, it is in a shared commitment to radically democratic ethics that such a potential lies. It is as a part of a substantial commitment to democracy that representations of the always wonderful, wild, and unknown world of the nonhumans can add to the irreducible diversity of a pluralist society. And it is when represented as part of the performance of a democratic citizen that humble understandings and relations with nature can be considered worthy of democratic consideration.

A commitment to democracy is a commitment to freedom and equality. There is no reason that commitment cannot be expanded to include nonhumans. The struggle comes in how that is accomplished, how freedom and equality are defined, and how nature can be included without being totally colonized. There are no simple answers to these issues. Therefore, there must be political discussion in order to share opinions concerning this and other questions that accompany the expansion of politics. This chapter has been an attempt to show where some of the potentials may lie.

Conclusion

A Feral Citizen's Democratic Imperative

Robert Ivie (2002, 281) declares that "an absence of dissenting voices in a democracy is the true sign of weakness and vulnerability, of a deep distrust of democracy and a failing faith in freedom, whereas speaking out is the patriotic duty of democratic citizenship." Castoriadis (1997a, 413) similarly claims, "A movement that would try to establish an autonomous society could not take place without a discussion and confrontation of proposals coming from various citizens." He follows this claim by exclaiming, "I am a citizen; I am formulating, therefore, my proposals." This book is one manifestation of my dissenting voice and my proposal as a citizen.

My offering to democratic culture comes in the form of a radically democratic approach to citizenship that relies on persistent wandering and disruptive engagement with others. Rather than offer a proposal that attempts to solve the many paradoxes of democracy, the theory and practice of feral citizenship that I offer is intended as a contribution to political agency and democratic culture. The purpose is to encourage a greater degree of political dissent; to stimulate political discussion; and to draw attention back onto the political sphere and its relevance vis-à-vis the proliferation of social movements, the loss of the markers of certainty, and the predominance of democratic regimes, discourses, and forms of legitimacy.

Given the strong and substantive commitment to democratic ethics, the point of the activity and methodology of feral citizenship is less to persuade than it is to incite and radicalize all those who rely on and use democratic terrain. Along with being committed to the antiauthoritarian and informal side of democratic ethics, the unaffiliated nature of feral citizens gives them the unique ability to focus their attention and activity on revitalizing the public sphere. The particular approach of feral citizenship is intended as a stimulant of political space and interaction that entices others to become political in their own way. I share Castoriadis's (1987, 2) belief that results presented as systematic and polished totalities are misguided. Similarly, I share with Arblaster (1987, 2) the belief that democracy is a perpetually contestable and critical concept that functions as a corrective to complacency. Sharing both traits, I thus defend feral citizens as critical political agents of democracy who struggle to create not a following but political moments that help revitalize public spheres that will, in turn, help draw attention to the need to recognize the value of the ethicopolitical attributes and conditions of democratic culture.

It is because healthy democracies require active citizens that wandering and feral are grafted onto citizenship. In fact, as I argued in the introduction to chapter 1, the reasons behind the articulation of the three terms is relatively simple. I support wandering because aimless travel is both pleasurable and essential; I support the feral qualifier because a feral's activity and presence is inherently disruptive; and I support citizenship because it is a common subject position that prioritizes the political and, in its informal capacity, can be at least potentially shared by all who live in a democratic society. Together these three descriptors offer a unique and radically democratic way of traveling what Arblaster calls the always-wanting democratic terrain of the present.

By drawing attention away from the relation to the state and by asking critical and democratically inspired questions, feral citizens create political moments that exist outside and not necessarily in relation to the typical dynamic among the state, the economy, and the media. The spaces created are genuinely political and intrinsically valuable as they are not intended to lead to anything other than a broadly revitalized public sphere. Feral citizens view politics as a terrain in which to wander rather than a legitimacy path along which to walk; yet they also view those using the paths in alternative ways as potential allies in their synagonistic battle against authoritarianism and the ever-expanding economy. By promoting and performing peripatetic aimlessness, feral citizens combine the pleasures of physical and theoretical wandering with the pleasure of exploratory politics.

Feral citizens are not trying to persuade others of the superiority of feral citizenship; they are, however, trying to show and perform the need for vibrant democratic culture and activity. Democracy, if it is to retain any of its liberatory and humanist potential, has to be more than the proliferation of individual social movements struggling for recognition, rights, and redistribution of goods. Necessary activities like those performed by social-movement activists will always be a significant part of a democratic society with an administrative body, but they must never be allowed to become the only part. In other words, while necessary and representative of a new stage in democracy's historic development, social movements are certainly not sufficient when it comes to embodying the tensions and struggles of democratic culture.

This relation between necessary and sufficient conditions is important for feral citizenship as it gives feral citizens the humility required in order to realize that along with all other approaches to political agency, feral citizenship is one part of a much grander and diverse struggle that has far more components to it than any one approach to citizenship could ever encompass. This is why part of the methodology of feral citizenship includes assisting others in becoming political in their own manner. The intention is not to win over but to politicize; it is this intention that allows feral citizens to view the proliferation of social movements as a particularly and/or potentially promising condition rather than a lamentable one that threatens the political sphere and should be resisted.

As a perpetually disruptive political agent with both wanderlust and a substantive commitment to democracy's critical and unfinished project, the feral citizen is not interested in the rebuilding of the disrupted social movements and communities, although s/he recognizes the need for others to do so. S/he will not necessarily be a welcome guest or achieve any of the glory associated with heroic acts. Furthermore, as the activities and disruptions s/he causes will not lead to any tangible results, the actions will always be difficult to value. This is why it is important that feral citizens choose their feral identity and realize the difficulties that accompany a commitment to feral citizenship. The virtuosity demanded of the feral citizen is not, however, as burdensome as it may at first appear. Indeed, for feral citizens who enjoy adventure and disruption, the virtuous requirements are joyful and the virtuosity may be best described as playful virtuosity. For feral citizens, playful engagement and disruption and the related commitment to democratic ethics and responsibilities are inseparable; for acting feral citizens, desire and duty are one and the same.

By the nature of their critical methodology, feral citizens create short-lived micropolitical moments that allow relations to redevelop in the context of political debate and opinion formation. On an individual basis, feral citizens are constantly changing political creatures who wander the democratic terrain with the desire to explore and understand the diversity, tensions, and hidden solidarities that litter democratic culture. But feral citizens are more than mere observers. Accompanying and encouraging observation and active listening is the desire to expand the political sphere, politicize relations between the state and social groups, and encourage social movement activists to realize their dependence on democratic culture. Thus, along with being active observers, feral citizens are also storytellers and actors. While intricately related and while all inform each other, the three political characteristics are separately essential to feral citizenship. Typically, when performing in public, political actors focus on the need to persuade others of the value of their particular concept of the good life. Feral citizens know they do not have the answer to democracy's many tensions, so along with being interested in revitalizing democratic culture, they are also perpetual learners who perform as active observers as well as narrators and unpredictable actors.

The identity as spectator or observer renders the feral citizen a constant learner always searching out more and more of the wonderfully diverse world. The search, however, does not make the feral citizen more of an expert, or more capable of speaking truthfully than others. What it does is make the citizen aware of the limitations of all political theory and practice and far more aware of her or his permanent amateur or apprentice status. Thus, as an active listener who is reliant on a plurality of places to visit and learn from, a feral citizen will never threaten the diversity that constitutes modern pluralist democracies. What feral citizens do is respond to the diversity by sharing stories about the diversity that exists, by creating more opportunities for the diversity to be valued, and by defending the intrinsic value of a pluralist society. It is as a storyteller that the defense of diversity first becomes political. When feral citizens wander and observe, they also engage. At this point, the understanding of the irreducible need for plurality becomes most apparent as it guides the democratically inspired storytelling. As a traveler, the feral citizen shares the many stories among the communities s/he visits. The sharing of stories helps uncover the diversity that exists; listening to them helps inform the critical and disruptive questions that feral citizens always pose to the visited community.

Finally, the identity as an actor allows the feral citizen either to perform as a part of the play taking place in the specific political moment or to create a new political moment that encourages others to perform and defend their positions. As an actor, the feral citizen focuses on enticing others to speak and defend their beliefs in a public forum. The feral citizen critically engages with these others not to destroy their beliefs or to get them to give up on their own activities, but to realize the situational context, the political promise, and the inevitable limit to all particular activities. The end result is ideally the shared realization that democracy is something worth protecting and the subsequent expansion of conversation and activity that addresses such a consensus.

Many years ago, Emma Goldman lamented the fall of socialism from its proud role as critic of instituted society to its reformist adjustments to the confines of the cage in which it found itself once it decided to enter the world of formal politics. "Socialism," Goldman (1972, 81) explains, once represented "the proud, uncompromising position of a revolutionary minority, fighting for fundamentals and undermining the strongholds of wealth and power." Now, "led astray by the evil spirit of politics," Socialism's only desire is "to adjust itself to the narrow confines of its cage" (1972, 80). While Goldman's polemics may seem exaggerated, the danger of the reformist cage is real. Many once radical and politically threatening social movements are now "legitimate" participants within the political and economic systems they once directly challenged. While Goldman does not distinguish between formal and informal politics, the political cage she speaks of is the formal political cage full of requirements and assumptions that force radical disruptors to become, at best, reformist disturbers and at worst, as is often the case with the desire to "allow" the subaltern to speak, mere performers of the imposed identity that accompanies the "opportunity" to speak.

Many, if not most, progressive social movements emerged as part of a passionate plea and demand for the expansion of the sphere of respect, recognition, and distribution. In fact, their legitimate demands continue to coalesce around liberal democracy's failure to come through on its promise of freedom, equality, and social justice. This is why I argue Arendt is mistaken in her assertion that the rise of the social sphere is necessarily a threat to the political sphere. Many social movements had, and still have, strong commitments to democratic ethics and desires, but are typically far too caught up in particular "necessary" struggles to realize the democratic terrain they rely on is beginning to become more and more difficult to traverse.

At present, what is missing from Western democratic culture is a political agent that can help create opportunities for the hidden foundational commitments of democracy to come to the surface and become valued topics of discussion and debate. Feral citizenship responds to this absence by embracing a radically democratic and purely political approach to political agency that focuses first and foremost on the expansion of the public sphere. The feral citizen is only ever one character within a much larger political play that feral citizens are always trying to expand and keep active.

Ernesto Laclau and Chantal Mouffe (1985, 160), who conceive of social movements as "an extension of the democratic revolution to a whole new series of social relations," offer a way of viewing the proliferation of particular social movements as a democratically inspiring situation. Feral citizenship shares their conviction that social movements have a great deal of political promise, but feral citizenship also acknowledges that many social movements, if left undisturbed also contribute to the demise of the political sphere. Part of what acting feral citizens do is stimulate moments where noninstrumental discussions concerning the danger of adding to the forces of tyranny can take place. In this way, feral citizenship can help to skew the potentiality of social movements in the direction of the expansion of the forces of freedom.

Social movements disrupted by feral citizens are not required to give up their necessary battles within the state, economy, and media dynamic. These are essential battles that must continue to be fought. Social movement advocates merely need to accept that while an important component of democratic culture, an ethos of protection and defense of movements is not enough. There also need to be distinct and noninstrumental political spaces and activities where the impositions of predetermined relations are not dominant, and the relevance of democratic culture can be discussed, defended, and acted upon.

Feral Citizenship's Particular Promise

The previous points were primarily made in chapters 1, 2, and 3 where I defended the political importance of wandering, feral(ness), and citizenship. In chapter 4, I explained that there is a legacy of disruption that contributes two important justifications for the wandering and feral metaphors to be attached to citizenship. The first relates to, and expands on, the need to attempt to actively listen to the insights of others. By clarifying the fact that the institutional and normative foundations of liberal democracy fail to adequately consider the disruptive and essentially democratic demands of the newly vocalized subaltern, postcolonial theorists point to the need for disturbance.

I argued, in fact, that many of the critical points emerging from postcolonial theory ought to be read as justifications for the expansion of the political sphere and the revitalization of the democratic tradition. Likewise, the ongoing struggle to resist the homogenizing potential of sisterhood feminism shows the need to keep the struggle between individual freedom and collective equality active.

When critiquing feminist sisterhood, bell hooks (1984), for example, challenges the feminist movement to take its claimed commitment to equality seriously. According to hooks, taking the commitment to equality seriously requires situating feminist movements in a larger, counterhegemonic movement that necessarily goes beyond sisterhood. What arguments like hooks's demand is a continued space for active listening and the sustained realization of the broader need to eliminate all forms of oppression, not just the one focused on by the particular movement. Her critical engagement with her "sisters" does not challenge the validity of feminism; rather it challenges those feminists who believe there is no longer any need for political discussion and coalition struggles. Hooks disrupts the often-presumed sisterhood without threatening to destroy feminism, and she challenges feminists without denying the need for feminism. Her commitment, like the feral citizen's commitment, is to an unachievable ideal that requires a constant struggle for genuine equality and freedom.

Feral citizenship does not solve any of the issues brought up by postcolonial or feminist critics of democracy. It offers an approach to political agency that focuses on the dual insight that they raise: that beliefs and norms should always be challenged, and that active listening is always imperfect. Thus, postcolonial theorists and difference feminists are comrades who offer a legacy of disruption in places not always part of a feral citizen's travels.

Against those who see Hannah Arendt as no longer relevant to modern pluralist times, I suggest that Arendt, like most other public realm theorists can remain influential if her political theory is hybridized and resituated. My particular attempt to reclaim Arendt is situated in a desire to respond to the complexities of the relationship among active listening, speech, and democracy. In hybridizing the Arendtian table and theater metaphors, the polis becomes a distinctively traveling polis constructed by feral citizens (and others) who engage with a diversity of residents along the democratic terrain. The traveling polis is not representative of a universal or inclusive polis; each one is relative to the broader collective expansion of the public sphere brought about by the collection of political moments that together represent and help create the informal public sphere. In addition, feral citizens loosen the rules of performance. In fact, feral citizens want the theater to be guerrilla-like, enticing all present to participate. In this new theater, the relation among actor, audience, and storyteller is disturbed and the formality of theater is bypassed in order to allow citizens to perform not what they are told to perform as or expected to perform as but what they wish to perform as. It is as a part of this informal street theater that actors defend their activities and question the assumptions that accompany their activities. While there is no director or set story line to follow, the presence of feral citizens helps to ensure that acting, debating, spectating, and storytelling are oriented to broad desires for antiauthoritarian democracy, a politics-first position, and the belief that the public sphere needs to be expanded.

The final three chapters of the book were pseudo examples of feral citizenship in action. Moved by a feral citizen's take on Emma Goldman's maxim that "the strongest bulwark of authority is uniformity; the least divergence from it is the greatest crime" (1972, 93), I traveled along the trails provided by public realm theorists and green political theorists. In chapter 5, rather than support a particular approach to revitalizing the public sphere, I defend a broad community of political tension. The general argument I make when addressing public realm theorists is that, while Habermas may be the most "reformist" of the public realm theorists, he nevertheless advocates for an important distinction within the public sphere that offers a helpful entry point into democratic theory.

The most important political lesson to be learned from Habermas lies in his clarification of the relationship between the purpose of communication and the conditions needed for particular kinds of communicative action to take place. For Habermas, weak public spheres, free from the need to decide, are responsible for creating public opinion, a task beyond the means of the parliamentary bodies, or strong public spheres, which are more centralized and much less diverse. In fact, according to Habermas (1996a, 307), strong public spheres are arranged prior to the need to decide and have particular rules to be followed for "justifying the selection of a problem and the choice among competing proposals for solving it." The structures are organized around the "cooperative solution of practical questions, including the negotiation of fair compromises." Their role is "less to do with becoming sensitive to new ways of looking at problems than with justifying the selection of a problem and the choice among competing proposals for solving it." The weak public sphere, however, has a creative rather than reactive function and must have networks and spaces that are open, fluid, and inclusive in order to embody a way of acting that resists organization as a whole, in order to allow public opinion to emerge from plural, antagonistic, and fragmented societies.

The main lesson I take from this line of Habermas's thought is the conviction that the weak and strong public spheres are distinct realms of interaction, and the weak public sphere is filled with wild democratic agents. For Habermas, the weak and strong public spheres have particular roles to play in relation to grander ideals of achieving rational consensus within a pluralist society. Feral citizens, of course, have no interest in limiting the creative interaction of weak public spheres to such a conditional purpose. In fact, Habermas's failure to recognize most of the liberatory potential of this sphere of interaction and creation is directly challenged by feral citizenship's contribution to democratic theory.

The discourses that constitute Habermas's weak public spheres are domesticated and disciplined by the always-present gaze and rules of the strong public sphere before citizens are allowed to initiate and take on their creative purpose. Nevertheless, the democratic potential of the Habermasian weak public sphere should be recognized, for within the justification of such uninstrumental deliberative spaces lies a much more liberatory potential. Furthermore, the contributions to public realm theory offered by Chantal Mouffe, Cornelius Castoriadis, and Hannah Arendt all require weak public spheres (or something similar to them) as access points for their own political involvement in contemporary political debates. Feral citizens create their own weak public spheres but there are no a priori conditions and no pre- or postpolitical desires influencing their creation. Feral citizens merely create moments where political discussions can occur.

So in chapter 5, while I show Habermas's deliberative democracy to be an improvement over typical liberal democracy and thus a welcome site for agonistic engagement, I also find agreement with Mouffe with respect to many of the faults of Habermasian politics. I found, however, no satisfactory answer to the question of how to live and act in a democratic society in Mouffe's agonistic pluralism. What I did find in Mouffe (and similarly in Arendt and Castoriadis) was a useful commitment to uncertainty and the need to embrace the tensions that evolve in and constitute the social and political sphere in a society responding to the "dissolution of the markers of certainty." Arendt's (1958, 197) argument that the attempt to do away with plurality is "tantamount to the abolition of the public realm itself" is convincing; it needs constantly to be brought to bear on public discourse. In concluding the chapter, I argued that all four of the public realm theorists have distinctly different definitions of what politics means, and depending upon the issue at hand, each has an important contribution to make to public realm theory and the revitalization of the public sphere.

Habermas, as the most reformist of the public realm theorists, sees politics as the procedural movement from public opinion to legitimate law. Mouffe views political action as the move from antagonistic relations to agonistic ones where the other is not seen as an enemy to destroy but as an adversary to debate with. For Arendt, politics is primarily a joyful act that emerges from independent agents gathering together and performing the (un)imaginable. And Castoriadis sees politics as the activity oriented toward creating an autonomous society and autonomous individuals. These four distinct visions of politics all have merit, and all address certain aspects of the political game. None, however, encompasses all political activity. That is why, along with defending feral citizenship, I defend a community of political tension.

My second deployment of feral citizenship occurs in chapters 6 and 7 where I use feral-inspired techniques to examine how the environmental movement has articulated green political thought with democratic theory. What I found was that many green political theorists articulate a type of citizenship that "is as likely to reinforce injustice as to undermine it" (Young 2001, 684). Specifically, environmentalists tend to call for a fixed articulation of green concerns and citizenship whereby environmental needs and deliberative citizenship are assumed rather than challenged. The articulation requires little more than a commitment to a priori green concerns, and an acceptance of current definitions of citizenship. Environmentalists interested in politicizing nature, here are inclined to add to a larger tendency in which it seems "at times as if debates about political legitimacy and the meaning of democracy have already been decided, with little or no room left for other, more participatory and egalitarian interpretations of democracy's central principles" (Keenan 2003, 2). I nevertheless still find political promise in the articulation primarily for the disruptive possibilities opened up by bringing representations of the nonhuman into the political sphere and by bringing political rules of the game into environmentalism.

These two final chapters show that the politicization of nature offers wonderfully disturbing and explicitly democratic potential that ought to be recognized and celebrated by those who wish to expand the public sphere and defend the priority of the political. Unfortunately, they also show that green political theory is subject to the same struggles that tend to limit the political promise of many movement-inspired politics. While theorists of green citizenship show clear intent to ground green politics in democratic discourse, there is an accompanying desire to use the legitimacy of democracy to attain predetermined green outcomes. Consequently, a particularly reductive view of democracy is attaining dominant status. Prominent green political theorists including John Dryzek, Hartley Dean, Val Plumwood, Andrew Dobson, Robyn Eckersley, Peter Christoff, John Barry, Robert Goodin, and Robert Brulle all support a procedural-deliberative approach to democracy, as they see it offering the best way to allow representation of nature access to political decision-making bodies.

The dominance of deliberative democracy and the assumptions that accompany deliberative approaches to democracy continue to obscure the truly liberatory potential of the *idea* of linking the extension of politics to representations of nature. Such a link or boundary-crossing partnership will necessarily be tension filled and disturbing and should be embraced as such. As it is impossible and undesirable to represent, know or manage a true nature we must continue to search for ways to discuss and talk about human relations with nature in ways that are nor subject to the false belief that to talk about an other requires knowing the other in some kind of intimate or accurate way. Therefore the always political struggle involves finding ways to discuss and talk about relations with and to nature outside the confines of a desire or need to know.

I argued that the democratic possibilities of the partnership of environmentalism and citizenship lie in extending a democratic gaze into the way in which "green" issues are created, the fallacy of accurate representation, and the conditions for inclusion in political public spheres. In response to those who may argue that environmental conditions are far too dire for the uncertainties of politics, I argued that if green political theorists are going to turn to citizenship they must take the political seriously. In addition to the weighty recognition that green citizenship should only ever be a part of the green public sphere, taking politics seriously also requires recognizing and accepting the baggage that comes along with using the explicitly political identity of the citizen. The tension between politicizing nature and democratic theory needs to be seen as a "genuine tension," one that does not lead to answers, greater legitimacy, or tension-free partnerships.

Environmentalists should therefore move from approaching democracy as a tool toward approaching democracy as a commitment to ethical ideals of freedom and equality. The skew in the partnership must be in the direction of democracy, for otherwise it is far too easy to fall into what Val Plumwood (2002) has referred to as an ecorepublic, where the desires for environmental sustainability trump democratic relations of freedom, equality, and autonomy.

In defending the political specificity of citizenship, I make a pointed claim for the protection of difference. I do not suggest one's identity as citizen is necessarily more important than another, only that it is different. Feral citizenship is not intended to replace other approaches to citizenship; it is not intended as a replacement to social movements or particular strategies; it also is not intended to be the only activity any particular individual is involved in. Feral citizenship offers a way of seeing value in many spheres of activity, ethos, and subject positions; and it attempts to create the conditions by which the valuation can occur. My defense of difference is related to my belief that a democratic society, culture, and political sphere, is indeed the best we can do.

By offering an approach to political agency that is primarily oriented toward creating political moments that have space for listening, discussing, and exploring the diversity of modern democracy for its own sake, I offer a radically democratic approach to citizenship that is genuinely committed to expanding the political sphere. By encouraging this kind of approach to citizenship, I invite individuals to participate in the expansion of the public sphere. This kind of humble, yet radically democratic approach to citizenship is absent from most public realm discourse. Indeed, many public realm theorists appear to have lost sight of the democratic struggle to expand the public sphere and create the conditions for the primacy of the political. Not only does feral citizenship make this democratic struggle central to political agency but it also encourages and entices others to do the same, not necessarily as feral citizens but as active agents of change. Thus, to public realm theory and the broad fields of democratic theory and citizenship studies as well as all those turning to democracy for legitimacy sake this book is a challenge to return to the task of prioritizing the political. It suggests the need for an expanded public sphere and invites all those who depend on democratic culture to engage and struggle with each other as only through such engagement will the irreducible need for democratic culture and the primacy of the political be realized.

CONCLUSION

The intent of this book has been to incite. As one would expect, there are no clear conclusions to make, no maps describing where to go next. If this book has made the world of politics appear more intriguing, if it leaves the reader with an urge to explore the political landscape, if readers wonder where the communities they belong to fit within democratic culture, if readers have found themselves wanting to disrupt the comforts of their own surroundings, or if the reader is beginning to realize the need for active listening as well as active deliberating, the book has achieved its goal. Now please, go forth and wander, disrupt and explore.

NOTES

SERIES EDITOR'S PREFACE

1. See *China's Environmental Crisis: Domestic and Global Political Impacts and Responses* edited by Joel Jay Kassiola and Sujian Guo (2010).
2. See Carl Boggs's *Ecology and Revolution: Global Crisis and the Political Challenge* (2012).
3. He does so in the expected feral way: critically, pointing out deficiencies in these theorists' works even though he accepts and relies upon other portions of their ideas.
4. Garside appears to recognize the resemblance of his constructed feral citizen to Socrates in an endnote to the Introduction (endnote 4) where Socrates's peripatetic methods are noted in connection with the feral citizens' wandering. However, the parallelism between Socrates and the feral citizen, as I see it, goes beyond this one dimension mentioned by Garside.

INTRODUCTION: DEMOCRACY AND THE FERAL CITIZEN

1. This argument is in the "Our Aims" part of the journal *Democracy and Nature* 2000–2005.
2. For Arendt, greater truthfulness or accuracy when one is practicing representative thinking comes from visiting others and representing the opinions of those others while deliberating.
3. Arendt would likely disagree with the liberatory potential that Laclau and Mouffe find in modernity. This disagreement does not, however, have any impact on similar interests in terrain, traveling, and trespassing.
4. The Athenian political philosophers were also wanderers, and peripatetic methods were certainly relevant to Socrates.

1 WHY WANDERING

1. The reduction of politics to decision making and walking to using treadmills is just one of the many parallels between representative

democracy and the treadmill. I will elaborate on other parallels in chapters 4 and 5.
2. Neil Evernden (1992, 120) has argued, "It is wildness that is destroyed in the very act of saving it. Wildness is not 'ours'—indeed, it is the one thing that can never be ours. It is self-willed, independent, and indifferent to our dictates and judgement. An entity with the quality of wildness is its own, and no other's. When domestication begins wildness ends." As a part of feral citizenship it is the wild that makes this citizen essentially unrepresentable.
3. Raphael Samuels suggested that "hiking was a major, if unofficial, component of the socialist lifestyle in the first three or four decades of the twentieth century" (Solnit 2000, 164).
4. One of the great ironies of the history of rural walking in England and elsewhere is that "a taste that began in aristocratic gardens should end up as an assault on private property as an absolute right and privilege" (Solnit 2000, 167).
5. Kinder Scout is the "highest and wildest point in the Peak District" and is the place of "the most famous battle for access" (Solnit 2000, 165). Up until 1836, Kinder Scout was public land. In 1836, "an enclosure act divided the land up amongst adjacent landowners, giving the lion's share to the Duke of Devonshire, owner of Chatsworth" (Solnit 2000, 165). Walkers called Kinder Scout "the forbidden mountain" as there were no footpaths that went anywhere near the summit. While there is a long and fascinating history of the battles that took place in order to gain access, it was in 1932 that the BWSF (British Workers Sports Federation) organized and publicized a mass trespass that while not supported by all Ramblers clubs "drew four hundred to the nearby town of Hayfield." Five of the trespassers were arrested, scuffles took place between the landowners and Ramblers, and the scene was set for a battle between Ramblers and landowners. Jail sentences between two and six months for "incitement to riotous assembly" were given to those arrested. The sentences outraged other Ramblers and members of the public, which brought both the "curious and the committed" to Kinder Scout. Ten thousand Ramblers showed up and "further mass trespasses and demonstrations were held in the wake of the verdict." As Solnit (2000, 166) explains, "The politics of walking heated up," and this battle was largely responsible for the creation of the Ramblers' Association, which helped create The National Parks and Access to the Countryside Act. This act specified that if there had ever been any right-of-way access points on one's property, they were open to the public.
6. As seemingly progressive as Britain has been toward wandering citizens, their reaction to Nomadic peoples like the Roma is far less encouraging. In 1994 the British Criminal Justice Act, for example, was amended to "abolish the duty of local authorities to provide sites for them." This was done in conjunction with increased "penalties for

setting up roadside camps" (Bauman 1998, 59). "The sites," Bauman explains, "are often inadequate and unhealthy, usually in undesirable locations, and always so stringently regulated that they cause a radical loss of freedom and deterioration in life-style" (Bauman 1998, 59). So through criminalization of their lifestyle, forced settlement in undesirable land, and consistent surveillance, the Roma as a nomadic group of travelers is gradually filtered out of the United Kingdom. Sadly, the way the Roma were treated in the post–cold war United Kingdom is better than elsewhere in Europe. For a fascinating while highly disturbing account of the way the Roma have been persecuted throughout history, see Janina Bauman's (1998) "Demons of Other People's Fear: The Plight of the Gypsies," referenced in the bibliography.
7. Guy Debord and the Situationists were very influential in the Paris uprising against de Gaulle in 1968.
8. Those who have written about the flâneur regularly refer to him as a hero of modernity.
9. His paintings were also intended to be the opposite of museum pieces. They were to be used and discarded or painted over.
10. Baudelaire (1863, 2) explains that "curiosity may be considered the starting point of his genius...Curiosity had become a compelling irresistible passion" and that the condition of convalescent, most like a return to childhood, helped the flâneur see "everything as a novelty...Nothing is more like what we call inspiration than the joy the child feels in drinking in shape and color."
11. Benjamin's description of the flâneur was less active and more leisurely than Baudelaire's, so I have chosen to emphasize Baudelaire.
12. Buck-Morss (1986, 111) suggests that "the flâneur in capitalist society is a fictional type; in fact, he is a type who writes fiction," but she sees this as a lamentable state because for her the flâneur ought to be much more. I believe the fictional approach could actually be a way of surviving in capitalist society while also retaining some of the independence and fantasy of the flâneur. What made Guys's art so important was that it was explicitly situational and modern. Perhaps we need to look a little more openly at where such modernity appears today as celebrants of the present may well allow us to see promise in places we no longer focus on. The other significant point was that the art was intended to be the opposite of the museum piece. It was intended to be public and available to the public.
13. Expressing oneself is obviously most important to the specifically political potential of flânerie.
14. It is possible that the flâneur's hatred of glory and praise of idleness could have clouded historians' perception of his role as a heroic figure of modernity.
15. Not all streets are "the proper grazing ground for the flâneur's imagination." The streets must be "wide enough so that 'hanging around,' 'stopping once in a while to look around,'" are possible. And perhaps

more importantly "there must be enough interest in the street and houses that flank it to allure those who have time and urge to hang around" (Bauman 1994, 147). Current urban planning, the dominance of the automobile, generalized distrust of others, laws that make loitering and wandering illegal, and city plans that discourage interaction and focus first and foremost on individual security, all make the opportunity for, and potential relevance of, flânerie much different now than it was in Baudelaire's or even Benjamin's time. I do not, however, as Bauman (1994) seems to suggest, believe it makes it entirely irrelevant.
16. Flâneurs had no interest in creating space for others.
17. This need is more regulative than real as the inclusive ideal is never entirely achievable. The limits of any inclusive ideal are discussed in chapter 3.
18. Tester (1994, 13) argues that "thanks to Benjamin, the flâneur is often seen as living and dying on the streets of Paris alone, so that any generalization of the figure and the activity would be historically questionable at best." If this is truly Benjamin's belief, my disagreement is obvious. I think flânerie as a way of interacting with others can be resituated in a new time, place, and language game. Like any subject of the past, as Benjamin certainly knows, the flâneur of today would be quite unlike the flâneur of the past.
19. Baudelaire (1863, 4) saw democracy much as Arendt saw mass democracy. He warned, "The rising tide of democracy, which spreads everywhere and reduces everything to the same level, is daily carrying away these last champions of human pride, and submerging, in the waters of oblivion, the last traces of these remarkable myrmidons." However, I think he remains an important voice defending the ethical rather than the procedural features of democracy.
20. Modern nonpolitical flânerie, if it exists at all, exists in shopping malls and places like Disneyland and the West Edmonton Mall. Buck-Morss explains that "in our own time, in the case of the flâneur, it is not his perceptive attitude which has been lost, but rather his marginality" (1986, 104).
21. Bauman's explanation of capital's expropriation of flânerie is an accurate, if perhaps overly conclusive, vision of the modern flâneur. He explains the expropriation is much like the pattern once practiced, with astounding success, by the modern factory upon the craftsman's right and capacity to set the purpose and the meaning of his productive labors. In the factory, the craftsman of yesterday was called, like before, to exercise his workmanship; only the designs he was to materialize and the values he was to incarnate were no more of his making. In the world of big and smaller Disneylands, yesterday's free-floating flâneur is called, like before, to wander aimlessly; only there is an aim in his aimlessness now, a function, a utility, a design—none of which is of his making (1994, 150).

22. "Capitalism has two ways of dealing with leisure, stigmatizing it within an ideology of unemployment, or taking it up into itself to make it profitable. The dividing line cuts between prosperity and suffering, and it makes a great deal of difference on which side one falls" (Buck-Morss 1986, 113).
23. One might even argue that the tone of the analysis has been male/masculine thus far.
24. George Sand's heroic efforts to overcome the obstacles to female flânerie offer ample support to the case that flânerie was indeed a male pastime (Gleber 1999, 173).
25. Buck-Morss (1986, 125) goes onto explain that dolls once considered "a way children learned the nurturing behaviour of adult social relations," through capitalist relations have now become "a training ground for learning reified" relations. "The goal of little girls now is to become a doll." As in, *"be a doll and..."*
26. Suggesting there were female flâneurs or flâneuses is not the same as suggesting it is not necessary to critique the maleness of flânerie. It is simply to show that women were not entirely excluded from so-called male flânerie.
27. George Sand is a pseudonym for Amantine-Aurore-Lucile Dudevant and was originally used by Sand as a means to gaining access to the literary world, which like the streets of Paris was not open to women.
28. While there are plenty of examples of Sand's wonderful and critical writing, the one that best describes her feelings toward the bourgeois class can be found in the essay "Letter to the Rich," published in 1848, the same year as *The Communist Manifesto*.
29. Her prolific writing is one of many parallels between Guys and Sand; others will become more apparent throughout this section.
30. I do not mean, in any way, to minimize the fact of the greater safety for men than for women—especially in certain areas of cities and at times of the day. I only want to argue that these actual/physical limitations should be fought against in order to allow women to be as comfortable with wandering as some men.
31. Situationists were a group of individuals who met periodically to discuss ways to bring Marxism up to date. They were very much interested in the politics of everyday life and finding ways of living life freely and fully. They resisted ideology of any sort and focused on what happened to people on a daily basis.
32. Bauman in particular, and also Buck-Morss and Benjamin, recognize and lament the degree to which capitalism and the market economy have co-opted the flâneur's once marginal way of dealing with life's realities.
33. The circular nature of this argument is purposeful as that is what makes the spectacle so difficult to focus on and/or challenge.
34. The integrated spectacle represents the most recent stage of the society of the spectacle where there is little if any space left for resistance

due to "incessant technological renewal; integration of state and economy; generalized secrecy; unanswerable lies; [and] an eternal present" (Debord 1998, 11).
35. Disinformation, unlike a simple lie, has to contain a degree of truth.
36. These unusual and unique few are not elites; they are independent thinkers within times where conformity and discipline are touted as desirable attributes.
37. The necessity of particular resistances is another manner in which the spectacle becomes more real. The society of the spectacle must create moments of activity but these moments are minor and solvable and thus give credibility to the spectacle and keep the attention away from the broader fantasy of spectoral democracy.
38. The simplest example of this occurs when we hear in the news comments like "here's what you said" and "the public believe." In such cases, the spectacle gives the appearance of agency and quickly takes it away by telling the spectator what s/he thought or thinks before the question can be considered.
39. The Situationists envisioned a genuine transformation of humanity and their art that included collages, written work that could be read in any direction, and films that had long moments of silence, and that intended not to express the passions of the old world but rather to invent new passions and extend life's boundaries (Jappe 1999, 65).
40. Sandilands (2003, 39) correctly points out that the flâneur's "ability to be both himself and an other, to both recognize and interrupt the conditions enabling his walking, makes him a powerful figure with whom to think through some of the sights, and sites, of late capitalism. His activity is both complicit with, and subversive of, the dominant relations of the market; thus his doubled view offers an experiential critique of commodity fetishism."
41. Chantal Mouffe is also convinced of the need to not abandon language simply because it has been taken up by an opponent. Language is situated but not fixed.
42. Once again, I must note that it is the difference between two kinds of activity that I am emphasizing. There is nothing to suggest that feral citizenship cannot emerge from movement spaces or that movements are essentially apolitical, there is, however, good reason to make sure the two kinds of activity are rendered distinct and accepted as essential components of democratic culture.
43. Solnit (2000, 55) reminds us that there are always exceptions or hybrids such as a woman known as Peace Pilgrim who set out "to remain a wanderer until mankind [sic] has learned the way of peace." Apparently, Peace Pilgrim kept walking for 28 years. In this sense, wandering feral citizens are almost like democratic pilgrims and they do not, like Peace Pilgrim, expect to achieve their goal. However, unlike Peace Pilgrim there is not such a divide between their wandering and their end desire. Feral citizens, as they wander

through the political terrain, are helping to constitute that which they desire.
44. A pilgrimage makes an appeal; a march makes a demand (Solnit 2000, 58).
45. Nelson Mandela's Long Walk to Freedom, Gandhi's tactical protest walks, and the Narmada valley marches to resist the development of dams in the valley to name a few of the best known, and the numerous less known but no less relevant, pilgrimages and marches that deserve much more attention than I am able to offer here.
46. Situationists were convinced that direct communication such as that practiced and promoted by feral citizens represented the only real way to banish social hierarchies and autonomous representations (Jappe 1999, 40). Habermas refers to this space as the essential opinion forming weak public sphere; for Chantal Mouffe, it is the irreducible tension-filled social sphere of agonistic democracy; for Hannah Arendt, it represents the essential condition of plurality that creates and nurtures the active condition of the *Vita Activa*; and for Castoriadis, such direct political interaction is necessary in order to approach anything resembling true democracy. For me, this family of resemblances points to the recognized importance of informal politics.

2 Why Feral

1. While it is feral animals and the manner in which such animals are treated that most influence my ideas on feral citizenship, there are also interesting examples of feral plants, which, like their animal counterparts, are often seen as a blight. Many ecologists seem to believe that tension-free existence, at least on a broad scale, is most important and best represents a healthy community. I have argued that when it comes to democracy, a lack of tension is a sign of trouble.
2. While there are certain parallels between an anarchic and a feral approach to citizenship, particularly in relation to their common allegiance to antiauthoritarianism, they are not identical. A feral citizen is not responsible to anarchist politics and ethics, and would be as disruptive among anarchist communities as all others would.
3. The reclaiming is from procedurally focused democrats who try to create conditions for consensus and harmony within the public sphere.
4. I am thinking particularly here of the many ways citizenship is being invoked by social movements who wish to use the legitimacy of citizenship without accepting the conditions that ought to accompany such use. There are obviously more extreme cases of (ab)use of citizenship that could be taken to task and need to be challenged. The reason I have chosen social movements as my primary target is because they are often progressive in their desires and represent

important counterhegemonic allies as far as striving toward increased freedom, equality, and social justice. My purpose is not to deny their relevance, but merely to ensure that many of the activities of social movements are recognized as distinctly different from the type of politics I claim is in decline and thus in need of revitalization. My purpose is also not to deny the relevance of those social movements for whom citizen rights—to refuge, to marriage, to choice, to bodily integrity—are the subject of their cause. I merely want to point to the importance of the fact that whatever rights are achieved there is always more to do when considered from the perspective of democratic citizenship.
5. Remarkably, it is assumed that purposeful eradication of the species by humans is more "natural" than an environments' adapting to the introduction of the new "invasive" species.
6. Of course, livestock are also not native to Missouri, but they are controlled and commodified and thus acceptable.
7. There are numerous other examples including an Australian web site (http://www.deh.gov.au/biodiversity/invasive/ferals/) for the Department of Environment and Heritage, which defines feral animals as mainly domestic animals gone wild and lists nine feral animals as invasive species, each with its own unnatural history and eradication plan.
8. In Bonnie Honig's book *Democracy and the Foreigner*, Dorothy from *The Wizard of Oz* is used as an example of a foreign founder who reinvigorates the community and then leaves "the people to sort out the terms of their own governance." Honig explains, "It is by virtue of her power as a stranger and a naïve that Dorothy can do what no native of Oz would dare to. Unsocialized by the reign of terror that has molded the locals into servile abjection, Dorothy topples the forces of corruption and alienation" (2001, 16). There are numerous connections between the claims that Honig makes about foreign founders and the importance of strangers to democracy and feral citizenship as an approach to political agency, for example, her claim that Dorothy's "final gift to the people of Oz is her own departure" (2001, 16).
9. In the case of Genie and Curtiss, the self-interest of Curtiss led Genie's mother to sue him for "supposedly slanderous exaggerations of the abuse Genie had suffered and for exploiting Genie for personal advancement" (Leiber 1997, 326), something I imagine would be very difficult for the researcher to resist.
10. Isolation is defined as absence from human contact and the distinction between being brought up by animals and being isolated apparently has no bearing on the information attained or attainable.
11. Along with actual cases, there are fascinating histories of human-animal relations that go as far back as Remus and Romulus all the way to Mowgli and the racially charged Tarzan (white) King of the

jungle. There are also the equally fascinating examples of hybrid creatures like Bigfoot, Yeti, Sasquatch, and Mono Grande, all of which disturb and disrupt notions of what it is that distinguishes humans from nonhumans. It is the disruptive influence that is most relevant to feral citizenship.
12. This hints at the conservative nature of normalized and institutionalized knowledge and why it is particularly important that the feral descriptor is *chosen*. Accompanying such a choice is a great deal of pressure and suspicion and the identity is neither for everyone nor permanent.
13. Known as Juliette, Madame Recamier was a French society leader, whose salon attracted Parisians from the leading literary and political circles of the early nineteenth century most of who at the time were fascinated with the ideas around Rousseau's Noble Savage.
14. Itard's housekeeper who cared for Victor, took him on pleasurable walks in the garden, and showed him the affection and love he craved (Newton 2002, 117).
15. Yousef (2001, 248) explains that careful examination of the case of Victor of Aveyron "leads to a more complex understanding of the state of nature than is suggested when the wild child and the man of nature are placed in strict opposition to one another." It also crosses the comforting line that has been drawn between human and nonhuman animals.
16. The Swedish taxonomist Linnaeus was so frustrated with where to fit in feral children that he actually "separated them on the pre-Darwinian biological tree as *Homo ferus*" (Steeves 2003, 244).
17. Yousef (2001, 256) who focuses on the eighteenth-century interest in feral children explains that "speculation regarding the significance of wild children for the study of human nature involved not only natural historians but also, and more importantly with regard to the question of education, epistemologists for whom the mind of the wild child could be seen as a real instance of purely hypothetical models of mental development." The discovery of a feral child has always brought early excitement and endless speculation over the potential secrets this child could reveal.
18. Equally curious is the focus on authenticity when it comes to feral children. There appears to be a real desire to disprove the truthfulness of the feral child's past but the specificity of the past matters little in relation to the knowledge that can be attained from the lack of communicable language and human interaction.
19. Along with Rousseau, there were other interests and writings in the late eighteenth and early nineteenth century that made the stirring around Victor the feral child of Aveyron even greater than in other cases. "Primitivism was in vogue, the subject of fictions by radical novelists such as Robert Bage (1720–1801), whose book *Hermsprong* (1796) depicted a hero who had been brought up among American

Indians before returning to the drawing rooms of England. William Wordsworth (1770–1850) and Samuel Taylor Colderidge (1772–1834) had only just published their *Lyrical Ballads* (1798) in which the insights of children, savages, and idiots were treated with revolutionary interests and respect" (Newton 2002, 107).
20. It should be noted that Rousseau's noble savage was carefully placed in a fertile and lush environment; abundance meant he did not need to consider the needs of survival, he could consider other endeavors.
21. One character that I have not discussed is the character of the Wandering Jew, who is the legendary figure who apparently mocked or mistreated Jesus on his way to the cross and was condemned to spend the rest of his life wandering the earth until the Judgment Day. The feral citizen is not condemned to wander, but finds joy in wandering and playfully disrupting her/his surrounding. So, along with the Wandering Jew being a highly controversial figure, there are important differences that make the connection far from obvious. Thus, I have not left the Wandering Jew out of my justification for wandering and feral citizenship because it does not add to my argument. On the contrary, I believe the legend could be read as further justification of my defense of disruptive wanderers. Rather, I left out the Wandering Jew as the story and the relation between the forced situation of the Wandering Jew and the chosen situation of feral citizenship requires much more in-depth examination and discussion than I am able to give in this book.

3 Why Citizenship

1. I do not mean to suggest that political agency should not be used by those with particular ends. On the contrary, I think it offers a terrific potential to expand the political sphere. What I argue, and why I choose citizenship as the appropriate identity, is citizenship includes an explicit commitment to democracy and the primacy of the political. I thus agree with Barber that "citizenship is not necessarily the highest or the best identity that an individual may assume, but it is the moral identity par excellence" (1984, 224), and it is the identity capable of becoming a supportive part of all other identities or subject positions that make up an individual. A partnership with citizenship ought to be recognized as a skewed partnership and the skew is always in the direction of democracy.
2. Mapping out and comparing the numerous ways democracy is used and abused by those who turn to it for legitimacy or guidance is beyond the capacity of this book. The difficulty of such a project lies in the lack of clarity that accompanies the use of the term and a rather unfortunate tendency to assume what we in the West are living in is democracy, and therefore, it does not need to be defined.

3. John Dryzek (2000, 160) adds to this argument by suggesting, "Democracy is, if nothing else, both an open-ended project and an essentially contested concept; indeed, if debates about the meaning of democracy did not occur in a society, we would hesitate to describe that society as truly democratic."
4. The absence of place, fit or expert knowledge renders the feral citizen a political amateur but the amateurism is not a preprofessional stage. Rather, it is a celebrated existence that helps stimulate the desire to explore, to wander, to create and travel on political terrain, and to engage with others while traveling.
5. Benjamin (1968, 159) explains that telling stories is one of the oldest forms of communication. He also notes that unlike other types of communication "it is not the object of the story to convey a happening *per se*," such as those exchanging "truthful" information. The storyteller is an active agent in the sharing of the story that is "passed on as an experience to those listening." The point to take from this is that storytelling is creative and active. A storyteller who is not merely a conveyor of information has agency and an active role in the translation. The goal is less accurate description or translation than it is creative performance of an experience. A storyteller is not a conveyor of truths s/he is a creative agent. If s/he is reduced to a conveyor of truth s/he loses the beauty of her performance and becomes "the disembodied conveyor of information," reduced in much the same way the modern citizen is reduced to a passive voter.
6. I am not suggesting here that social movements only struggle for recognition and redistribution of rights, only that this often-dominant aspect of their activity not be allowed to hinder them from joining other more explicitly political spaces.

4 Feral Citizenship as Method and Feral Citizen as Guide

1. Feyerabend (1978, 45) argues that "the semblance of absolute truth *is nothing but the result of an absolute conformism.*" Such an argument does not mean the accepted "truth" is necessarily "wrong"; it merely means it has gained common acceptance, which should never be thought of as a sign of truthfulness.
2. The politics of address is something far more integral to the thoughts of those on the margins than it is to those in the center. Those on the margin often have to become like the center to be recognized. An active listener who acknowledges the power in the politics of address must accept that s/he is not always the one being addressed. They need to learn how to hear differently and address others differently. At times they will need to listen in and at other times they may be able to participate; their responsibility is to find

out how and when to perform each part. This is no easy task and there is no obvious answer. What there is, is a need for active listening. Once active listening is acknowledged as an essential part of democratic speech acts it becomes apparent that *more* active listening is always necessary particularly among those who occupy spaces of power. Those within dominant culture must realize that not all speech addresses them and not all speech acts involve their direct participation.
3. Significantly, Fanon's desires require participation. If one desires to go for a walk that desire is not satiable by someone else walking on your behalf. Desires require participation and thus must be acted upon. If democracy is desirable, then, it is not something that ought to be left to representatives who act (or work) on behalf of desiring subjects.
4. As Anzaldúa says elsewhere when discussing the relationship of mestizas with white society, "they will come to see that they are not helping us but following our lead" (1990, 384).
5. This, of course, does, at times, require something from those in positions of authority but that discussion is other than the disruptive and political one that is initiated by Anzaldúa's demands to accept her otherness as worthy of recognition and consideration.
6. The "you" referred to here is the colonist but could be extended to all interlocutors.
7. Mignolo (2000, 736–37) eloquently describes the situation, "Inclusion doesn't seem to be the solution to cosmopolitanism any longer, insofar as it presupposes that the agency that establishes the inclusion is itself beyond inclusion, 'he' being already within the frame from which it is possible to think 'inclusion.' Today, silenced and marginalized voices are bringing themselves into the conversation of cosmopolitan projects, rather than waiting to be included. Inclusion is always a reformative project. Bringing themselves into the conversation is a transformative project that takes the form of border thinking or border epistemology—that is, the alternative to separatism is border thinking, the recognition and transformation of the hegemonic imaginary from perspectives of people in subaltern positions. Border thinking becomes a 'tool' of critical cosmopolitanism."
8. There is a particularly interesting legacy of this sort of persuasive disruption within feminist "we" spaces that is far from solved with third-wave feminism and has led, most recently, to an interest in drawing connections between feminists and cosmopolitans. Believing *cosmofeminism* might help draw other "universalisms into a broader debate based on their own situatedness," Pollock, Bhabha, Breckenridge and Chakrabarty (2000, 584), for example, suggest to succeed it would need to "create a critically engaged space that is not a screen for globalization or an antidote to nationalism but is rather a focus on projects of the intimate sphere conceived as part of the

cosmopolitan." Hooks, however, focuses her critique more on feminism and its relation to gender issues.
9. For Braidotti (1999, 91) figuration "is a way of bringing into representation the unrepresentable, insofar as it requires awareness of the limitations as well as the specificity of one's locations. Figurations thus act as the spotlight that illuminates aspects of one's practice which were blind spots before."
10. The full definition that Haraway gives is nearly 20-pages long and thus far too long to include here. For those interested, it is titled "'Gender' for a Marxist Dictionary, The Sexual Politics of a Word" and is in Haraway's (1991) collection of essays *Simians, Cyborgs, and Women: The Reinvention of Nature*.
11. I believe, with Young (1997b, 360), that Haraway's situated objectivity is close to Arendt's enlarged thought. If it is true that for Arendt a plurality of perspectives must be maintained for publicity to be preserved, then the idea of a view from nowhere attained through transcending all particular points of view is contradictory.
12. When Spivak discusses representation and the subaltern she distinguishes between representation as speaking for and representation as speaking about. The former is political representation while the latter is more re-presenting "in the sense of making a portrait" (Kapoor 2004, 628). The conflation of these two meanings has led to dominant culture speaking for what they have re-presented and thereby silencing and recreating, in their own image, what they intended to represent. An always changing and mysterious identity like Haraway's disrupts both understandings of representation and exists outside the parameters of representation.
13. For the ancient citizen of Athens "politics was an inexhaustible, everyday 'curriculum' for intellectual, ethical, and personal growth" (Bookchin 1987, 60). For the feral citizen, the numerous social movements, political theories, and opinionated actors that populate the cosmopolitan terrain represent the ever-changing democratic curriculum.

5 Public Realm Theory, from State to State of Being/Becoming

1. The success can be measured by the lack of genuine or revolutionary alternatives to "radicalizing" liberal democracy even among those who state they are interested in revitalizing socialism and the Leftist tradition.
2. The key to understanding and exploring the political import of Jürgen Habermas's public realm theory is in recognizing his particular reformist focus on rationalization, decision making, and political legitimacy. These three foci are grounded respectively in,

universalization (rules of argumentation that act as bridging principles to make agreement possible); discourse ethics (a moral theory justified through its relation to universalization and rational lifeworlds); and the ideal speech situation (required for making the outcome of discourse ethics valid).
3. An ideal speech situation requires the establishment of "a network of pragmatic considerations, compromises, and discourses of self understanding and of justice, [that] grounds the presumption that reasonable or fair results are obtained in so far as the flow of relevant information and its proper handling have not been obstructed" (Habermas 1996a, 296).
4. In Mouffe's (2005) recent book *On the Political*, she focuses her criticism on Third Way and cosmopolitan approaches to politics. While the opponent or antagonist has changed, her key criticism remains the same, to understand politics in contemporary times one must understand the irreducibility of antagonism and accept that, far from being a problem for democracy, it is the foundation of democracy. Mouffe is critical of all approaches that look to eliminate conflict in order to achieve some sort of consensus.
5. There is a clear Western bias to the assumption of pluralism and the assumed allegiance to (liberal) democracy. I address this issue in chapter 2 by arguing for active listening and the need for continuous revisiting of assumptions such as the assumed allegiance and superiority of liberal democracy.
6. The irony lies in the fact that radical democracy may not be very radical at all as it continues to work within the confines of liberal democratic theory and practice.
7. It is generally acknowledged that Habermas is the theorist most responsible for reinvigorating the discourse around public realm theory (Benhabib 1992; Calhoun 1992; Villa 1992; Coles 2000; Mouffe 2000; etc.).
8. In response to the loss of social cohesiveness in one public sphere Habermas (in Calhoun 1992, 424) now accepts that "it is wrong to speak of one single public even if we assume that a certain homogeneity of the bourgeois public enabled the conflicting parties to consider their class interests, which underneath all differentiation was nevertheless ultimately the same, as the basis for a consensus attainable at least in principle."
9. Challenges to reductions of democratic citizenship to voting, or in the so-called participatory democratic approaches that call for increasing plebiscites and referendums, are obvious. But the challenges to those promoting direct forms of democracy are also significant. Participation in the decision-making stage does not make for a direct democracy. On the contrary, it allows for a more inclusive management decision that imposes responsibility on a populace that is considered responsible for making a choice, which is their

democratic duty. As Habermas shows, democracy occurs long before the decision-making process, and the imposition of having to attain democratic agency by answering a yes/no question considered appropriate by the unrepresentative state should never be accepted.
10. As a result of this belief Mouffe can make the seemingly anti-Arendtian argument that "it is impossible to determine a priori what is social and what is political independently of any contextual reference" (2005, 17). But again, the difference is more in kind than it is fundamental.
11. Presently it will become apparent, at least on this point, that there are similarities to this distinction that Mouffe makes and the distinction Habermas makes between weak and strong public spheres.
12. Mouffe (1993, 3) argues that the disappearance of the opposition between democracy and totalitarianism opens up the potential for the "establishment of new political frontiers" that could lead to numerous new friend-enemy relations that could either be taken up politically as ways of extending the democratic project, or devolve into a revitalization of essentialist and fixed oppositional positions.
13. There are two connotations of the word "imagination" when used by Castoriadis. The first connects the word to images in the general sense; the second is in relation to creation and invention (1997a, 321). Castoriadis asks his readers to consciously register the imaginary we are in while illuminating imaginaries that better serve democratic ends. The imaginary is never merely a reflection of a given reality; it is a situated and social-historical genesis of specific symbols, images, forms, and institutions. Every society has unique imaginaries that distinguishes it from other societies and is locatable in history as a "creation and ontological genesis in and through individuals' doing and responding/saying" (Castoriadis 1987, 3-4).
14. "Every symbolism," Castoriadis (1987, 121) explains, "is built on the ruins of earlier symbolic edifices and uses their materials...By its virtually unlimited natural and historical connections, the signifier always goes beyond a strict attachment to a precise signified and can lead to completely unexpected realms." I believe this unexpected potential has a great deal in common with Arendt's notion of natality, which she describes as the uniquely human "capacity of beginning something anew, that is, of acting" (1958, 11).
15. "Genuine questions," for Castoriadis, are questions that do not lead to end points but rather to questions of justice and freedom, questions that have no "legitimate" answers. "If a full and certain knowledge (*episteme*) of the human domain were possible," he argues (1997a, 274), "politics would immediately come to an end, and democracy would be both impossible and absurd, democracy implies that all citizens have the possibility of attaining a correct *doxa* and that nobody possesses an *episteme* of things political." Posing genuine questions would require a radical rethinking of democracy and its promise.

6 A Tough Walk: Environmentalists on Democratic Terrain

1. The human-centered view has created a fair amount of resistance among what may now be called the followers of classical eco- or biocentric approaches to human/nature relations. For an exemplary piece see Patrick Curry's less-than-favorable review of John Barry's "Rethinking Green Politics" in *Environmental Values* (2000, 120–22).
2. There are different terms used by contributors to this discourse, earth citizen, environmental citizenship, ecological citizenship, and green citizenship. I have chosen to use green citizenship in order to minimize the potential of reverting back to the overly simple distinctions between environmental and ecological thought whereby ecological is considered radical or deep and environmentalism is considered reformist or shallow (although "green" remains present in these distinctions).
3. Interestingly, there was a similar turn to individual responsibility in the early 1990s as a response to the Rio conference. Books with titles such as *Ten Things You Can Do to Save the World* and *Caring for the Earth* were all helpful in focusing responsibility on individual actors and taking it off governments and businesses.
4. Dean actually disagrees with the need for this disarticulation as he turns to deliberation rather than opinion formation and antagonism for green entrance into political communities.
5. The problem with Torgerson's green public sphere is that it is "green" rather than democratic and it could very easily bracket out other positions that were not green. Basically, it threatens to bracket out or limit the possibility for the spontaneous emergence of the unexpected.
6. Sachs has argued that much of contemporary environmentalism has led "capital, bureaucracy and science—the venerable trinity of western modernization—[to] declare themselves indispensable in the new crisis" by promising "to prevent the worst through better engineering, integrated planning and more sophisticated models" (Sachs 1999, 67).
7. It is fair to assume that the differences between the two theorists would be differences of kind rather than substance, and their disagreements would be between "friendly enemies" who would recognize the merit of each other's position.
8. While situating what counts as public may not, at first, seem like a concern for green citizenship, its relevance soon becomes apparent when the type of language used to make representations of nature politically relevant becomes the focus of attention.
9. Arendt (1968, 42) believes that one thing that renders humans mortal is that they are not merely a species. They are a plurality of distinct individuals each one with a potential to offer something new

to the world. Viewing humans as a species is like viewing them as a mass, easily duped, easily manipulated, and very sheepish. It is only by seeing humans as a controllable species that behavioral techniques and centralized authority can fathom a responsive-responsible mass of citizens behaving as they are told.

7 THE OBSCURED PROMISE OF GREEN CITIZENSHIP

1. MacGregor (2004, 78) correctly points out that "while green politics tends to question the boundary between public and private in terms of the obligations and duties of citizens, there is scant recognition that what takes place in the private sphere is much more than consumption and reproduction."
2. It is worth noting, however, that Dobson's move away from the focus on rights is not where the threat to democracy lies. Indeed his intent to distinguish himself from van Steenbergen (1994), and Newby (1996), both of whom wish to add ecological citizenship to T. H. Marshall's "three-fold typology of citizenship (civil, political, and social)," is commendable.
3. Insofar as women are seen as household consumers and carers of the home, they become the focus of Dobson's expanded sphere of responsibility. Far from assisting in liberating women from the burdens of often oppressive and disciplinary household relations, ecological citizenship runs the risk of reversing the trend by *increasing* dutiful responsibilities and imposing more rules of conduct.
4. Added to the danger is that fact that along with Dobson, Mark Smith (1998), John Urry (1998), Hartley Dean (2001), Deane Curtin (1999), John Barry (1999), and others all agree that the introduction of ecology to citizenship discourse is important for the challenge it represents to the rigid split between public and private realms of action. Smith (1998, 98) and Urry (1998, 12) actually believe the split cannot hold up in times where relations of entitlement and obligation are new.
5. Far too many green political theorists have been influenced by Robert Goodin's (1992, 168) claim that "to advocate democracy is to advocate procedures, [while] to advocate environmentalism is to advocate substantive outcomes."
6. Obviously, ecofeminism (which has its own authoritarianism that needs to be addressed) and the environmental justice movement have offered a great deal on this front, but they are largely absent from green citizenship discourse.
7. There is an increased potential of direct democracy when democracy is no longer defined as a decision-making procedure but rather the communicative interaction of publicly engaged citizens striving

toward understanding each other's position concerning the public good.
8. Barry (1999, 234) argues that "as a practice, green citizenship is the ethical core of collective ecological management."
9. Christoff (1996, 160) also believes that while ecological citizenship is transnational, "the state remains an exceptionally important focus of concern for ecological citizens and their organizations seeking to refashion its activities; to have it enshrine protection or generalizable environmental interests in legislation guaranteeing environmental standards, the protection of ecosystem species; and to provide the legal material support for further (ecological) democratization."
10. Mark Smith (1998, 82) actually takes the idea of an ecological contract one step further by suggesting, "What is needed is a 'greenprint,' a strategic orientation which acknowledges the complexity and uncertainty of the changing relationships between society and nature."
11. At no point are the economy, capitalism, hierarchy, domination, and so on directly challenged by those supporting the stewardship ethic.
12. John Dryzek (2000, 142) argues that any state operating in the context of a capitalist market system "is highly constrained in the terms of the kinds of policies it can pursue...policies that damage business profitability—or are even perceived as likely to damage business profitability—are automatically punished by the recoil of the market."
13. Newby's (1996, 219) question "can the tragedy of the global commons be avoided only by the advent of a modern day green Leviathan? Or can human stewardship over the commons be exercised by more democratically accountable needs?" shows exactly how one can be caged by the entry position.
14. Included in this list would be Murray Bookchin, Takis Fotopoulos, Catriona Sandilands, Douglas Torgerson, John Dryzek, Robyn Eckersley, and Val Plumwood. The latter three are key contributors to the section on ecodeliberative democracy.
15. This is, of course, unless the environmentalist is also a citizen in which case the responsibility is to the citizen and not necessarily the environmental position which is to be respected but not necessarily accepted. And, the citizen with the environmental position is to respect the voices of the others by actively listening and accepting the limitations of the environmental position.
16. It is worth noting that Cornelius Castoriadis and Hannah Arendt both see the modernist move from politics to ethics as one of the greatest threats to the revitalization of the public sphere
17. One can only care for what already exists. Citizenship, if it is to be about freedom, requires creativity, spontaneity, and unpredictable outcomes.
18. There is no reason why "refusing to speak transparently *from a place*," will necessarily "look hypocritical, unexamined, and neo-colonial"

(Curtin 1999, 169). A placeless position can, if recognized as such, be one of many voices in a pluralist ethic. Forcing one to create a sense of place or become native is not an amoral position and should be viewed with suspicion. Blind faith in local community is a dangerous move that itself should not be left unexamined. There is also the issue of immigration that is totally obscured for view in this place-based focus.
19. As if care were harmonious!
20. The villages were entirely inaccessible by road and in order for one to reach it even by foot travelers were required to "slide from root to branch along a slippery slope, in danger at every moment of losing control" (Curtin 1999, 171).
21. Of course, once placed in a polyvocal debate the villager's argument around rightness would be open to debate.
22. Listening needs to be to nonhuman nature and to other individuals who represent nature.
23. It may in actual fact be much more useful to focus on emancipation rather than freedom when we speak about politicizing nature.
24. That caring emerged from the household is significant, not because it was "women's work" but because it was assumed that those who cared for the children were capable of knowing what was best for them. There was, in ideal situations, a compassionate and well-understood relationship that was not democratic but caring. The carers' allegiance was to the children and there is little doubt that protection of the household and those within it would prevail over all other concerns. Nor did it matter what the children wanted for they were considered incapable of making rational choices.
25. The ethic of care does not engage in the necessary translation of the spirit of engagement as it continues to emphasize the personalized-care relationship rather than the public-justice relationship.
26. While I am borrowing Arendtian language here, it should be noted that Arendt would not consider nonhumans as capable of becoming whos. Against Arendt, I do believe nonhuman animals have the capacity to be individuals or at least not only adapt to their surroundings or their inherent nature.
27. Importantly, the nonhumans can also be a cohabitant and friend, but these relations are not the ones to be politicized (unless translated) as they are far more personal.
28. To live and dwell on the land is certainly not a requirement for the sort of citizenship I am talking about. It is, however, a difference and *can be* an important stimulus.
29. I will not spend time weighing the merits of their methodology as it is only the broad abstract finding of their research that is relevant to this paper.
30. Their discovery of the significance of walking came from ethnographic work they undertook in western North Carolina from 1996

to 1998. What was distinct about their work is that they interviewed not only self-identified environmentalists but "mountain residents" who are often seen (by self-proclaimed environmentalists) as the opposite of an environmentalist as they have the nerve to actually make a living off the land rather than visit it on weekend getaways in the SUV.
31. Thoreau also explains that a walk, both physically and spiritually, requires leaving familiar or ordinary life in order to venture into what he calls "the springs of life," an argument not at all unlike Arendt's claims for natality within always wild and wonderful political spaces.
32. Abstracting this theory, we can say that it is theoretical joyful walking like this that can help the actor and the antagonists to minimize the chance of any one identity gaining universal status related to a belief that they have "got it right."
33. This remains the case as long as the stories are not themselves translated into unchallengeable foundational truths.
34. It is important to be careful not to interpret this argument as a reification of local knowledge as authentic or superior. The argument merely supports the expansion of the amount of acceptable stories and the need to be careful not to silence those voices that have something to offer and must be listened to.
35. Eckersley argued that earlier authors from the Frankfurt school, especially Herbert Marcuse, were better partners for the goal of reconciling humanity with nature.
36. To create communicative rationality, discourse must be oriented toward consensus, allow for the unforced force of the better argument, and include all participants affected or potentially affected by the outcome.
37. Andrew Whitworth (2001, 34) goes as far as to argue that "not only *can* environmental discourses be judged by Habermas's normative standards—they *must* be" (emphasis original).
38. The oft-used example for supporting this position is the case of the baby who while incapable of participation is nevertheless advocated for in political deliberation. However, much like my example with caring for children in the household there is a sense of knowing much greater among adults and children than there is among humans and nonhumans.
39. Dryzek (2000, 154) argues, "Even if nature cannot receive equality in the politics of presence, it can receive such treatment when it comes to the politics of ideas," which is precisely what communicative ethics offers.
40. Plumwood (2002, 239) also argues, "The ecological message, no matter how persuasive to people at large, will never change policy while this is made by ruling elites who have a powerful stake in keeping the systems we have to change."

NOTES

41. Plumwood (2002, 72) speaks of many different kinds of remoteness: spatial remoteness, which would involve living far away from the consequences of decisions; consequential remoteness, where decision makers are unaffected by consequences; communicative or epistemic remoteness, which means an inability to communicate with those affected; temporal remoteness, which includes being remote from future effects of decisions; and finally, technological remoteness, which includes externalizing consequences of present luxuries.
42. While I do not make the argument here, it is certainly true that remoteness can also be used as a critique of the state and a critique of the cosmopolitan position.
43. I believe the key difference lies in listening to know and uncover nature's secrets and listening to hear and appreciate it for its own sake, for the beauty of its wildness, its unknowability, and its otherness.
44. There is no doubt that those who speak for nature speak in different voices. This can be seen even among activists particularly the animal right versus environmentalist voices where the former tends toward individualism while the latter tends toward collectivity.

Bibliography

Abensour, Miguel (2002). "Savage Democracy and Principles of Anarchy." *Philosophy and Social Criticism* 28(6), 703–26.
Adams, Katherine (2002). "At the Table with Arendt: Toward a Self-Interested Practice of Coalition Discourse." *Hypatia* 17(1), 1–33.
Adams, Kathleen (1985). *The Wild Women: An Inquiry into the Anthropology of an Idea*. Rochester, VT: Schenkman Books.
Agyeman, Julian and Bob Evans (2006). "Justice, Governance, and Sustainability: Perspectives on Environmental Citizenship from North America and Europe. In Andrew Dobson and Derek Bell (Eds.). *Environmental Citizenship* Cambridge: MIT Press, pp. 185–206.
Anzaldúa, Gloria (Ed.) (1990). *Making Face, Making Soul: Creative and Critical Perspectives by Feminists of Color*. San Francisco: Aunt Lute Press.
——— (1987). *Borderlands/La Frontera: The New Mestiza*. San Francisco: Aunt Lute Press.
Anzaldúa, Gloria, and Cherríe L. Moraga (Eds.) (2002). *This Bridge Called My Back: Writings by Radical Women of Color*. Berkeley, CA: Third Woman Press.
Arblaster, Anthony (1987). *Democracy*. Minneapolis, MN: University of Minnesota Press.
Arendt, Hannah (2005). *The Promise of Politics*. New York: Schocken Books.
——— (1990). "Philosophy and Politics." *Social Research* 57(1), 5–103.
——— (1979). "On Hannah Arendt." In Melvyn A. Hill (Ed.). *Hannah Arendt: The Recovery of the Public World*. New York: St. Martin's Press, pp. 301–39.
——— (1968 [1954]). *Between Past and Future: Eight Exercises in Political Thought*. New York: The Viking Press.
——— (1963). *On Revolution*. New York: The Viking Press.
——— (1958). *The Human Condition*. Chicago: Anchor Books Edition.
Arias-Maldanado, Manuel (2001). "Sustainability and Democracy: Towards a Green Democratic Model." *OCEES Research Paper No. 21*. Oxford, UK: Mansfield College.
Asen, Robert (2002). "Imagining in the Public Sphere." *Philosophy and Rhetoric* 35, 345–67.
Austin, Mary (1907). "The Walking Woman." *Atlantic Monthly* (100), 216–20.
Barber, Benjamin (1984). *Strong Democracy: Participatory Politics for a New Age*. Berkeley, CA: University of California Press.

Barry, John (2006). "Resistance in Fertile: From Environmental to Sustainability Citizenship." In Andrew Dobson and Derek Bell (Eds.). *Environmental Citizenship*. Cambridge, MA: MIT Press, 21–48.
——— (1999). *Rethinking Green Politics*. New York: Sage Publications.
——— (1996). "Sustainability, Political Judgement and Citizenship: Connecting Green Politics and Democracy." In Brian Doharty and Marius de Geus (Eds.). *Democracy and Green Political Thought*. London: Routledge, pp. 115–31.
Bartsch, I., DiPalma, C., & Sells, L. (2001). "Witnessing the Postmodern Jeremiad: (Mis)undelstanding Donna Haraways method of Inquily." *Configurations* 9, 121–164.
Baudelaire, Charles (1863). "The Painter of Modern Life." Accessed on October 13, 2004, available at http://www.idst.vt.edu/modernworld/d/Baudelaire.html.
Bauman, Janina (1998). "Demons of Other People's Fear: The Plight of the Gypsies." *Thesis Eleven* 54, 51–62.
Bauman, Zygmunt (1999). *In Search of Politics*. Stanford, CA: Stanford University Press.
——— (1994) "Desert Spectacular." In Keith Tester (Ed) *The Flâneur*. New York: Routledge, pp. 138–57.
Beck, Ulrich (1992). *Risk Society: Towards a New Modernity*. London: Sage.
Benhabib, Seyla (1999). "Citizens, Residents, and Aliens in a Changing World: Political Membership in the Global Era." *Social Research* 66(Fall), 709–43.
——— (1996a). "The Democratic Moment and the Problem of Difference." In Seyla Benhabib (Ed.). *Democracy and Difference: Contesting the Boundaries of the Political*. Princeton, NJ: Princeton University Press, pp. 3–18.
——— (1996b). "Toward a Deliberative Model of Democratic Legitimacy." In Seyla Benhabib (Ed.). *Democracy and Difference: Contesting the Boundaries of the Political*. Princeton, NJ: Princeton University Press, pp. 67–94.
——— (1992). *Situating the Self: Gender, Community and Postmodernism in Contemporary Ethics*. Cambridge, MA: MIT Press.
Benhabib, Seyla, Judith Butler, Drucilla Cornell, and Nancy Fraser (1995). *Feminist Contentions: A Philosophical Exchange*. London: Routledge.
Benjamin, Walter (1968). *Illuminations: Essays and Reflections*. New York: Schocken Books.
Berger, Johannes (1991). "The Linguistification of the Sacred and the Delinguistification of the Economy." In A. Honneth and H. Joas (Eds.). *Communicative Action: Essays on Jürgen Habermas's the Theory of Communicative Action*. Cambridge, MA: MIT Press, pp. 165–80.
Bernheimer, Richard (1952). *Wild Men in the Middle Ages: A Study in Art, Sentiment and Demonology*. New York: Octagon Books.
Bey, Hakim (1991). *T.A.Z. The Temporary Autonomous Zone, Ontological Anarchy, Poetic Terrorism*. Brooklyn, NY: Autonomedia.
Bhabha, Homi K. (1994). *The Location of Culture*. London: Routledge.

Bickford, Susan (1996). *Listening, Conflict, and Citizenship: The Dissonance of Democracy*. Ithaca, NY: Cornell University Press.
Birch, Thomas (1995). "The Incarceration of Wildness: Wilderness Areas as Prisons." In Max Oelschlaeger (Ed.). *Postmodern Environmental Ethics*. New York: SUNY Press, pp. 137–62.
Boggs, Carl E. (2012). *Ecology and Revolution: Global Crisis and the Political Challenge*. New York: Palgrave Macmillan.
Bohman, James (1996). "Deliberative Democracy and Effective Social Freedom: Capabilities, Resources, and Opportunities." In James Bohman and William Rehg (Eds.). *Deliberative Democracy: Essays on Reason and Politics*. Cambridge, MA: MIT Press, pp. 321–48.
Bookchin, Murray (1995a). "Communalism: The Democratic Dimension of Anarchism." *Democracy and Nature* 3(2), 1–17.
―― (1995b). *Social Anarchism or Lifestyle Anarchism: An Unbridgeable Chasm*. San Francisco: AK Press.
―― (1994). *Which Way for the Ecology Movement? Essays by Murray Bookchin*. San Francisco: AK Press.
―― (1987). *The Rise of Urbanization and the Decline of Citizenship*. San Francisco: Sierra Club Books.
―― (1971). *Post-Scarcity Anarchism*. Montreal: Black Rose Books.
Brady, Jon S. (2004). "No Contest? Assessing the Agonistic Critiques of Jürgen Habermas's Theory of the Public Sphere." *Philosophy and Social Criticism* 30, 331–54.
Braidotti, Rosi (1999). "Response to Dick Pels." *Theory, Culture & Society* 16(1), 87–93.
―― (1994). *Nomadic Subjects*. New York: Colombia University Press.
Brockelman, Thomas (2003). "The Failure of the Radical Democratic Imaginary: Žižek versus Laclau and Mouffe on Vestigial Utopia." *Philosophy and Social Criticism* 29(2), 183–208.
Brulle, Robert J. (2002). "Habermas and Green Political Thought: Two Roads Converging." *Environmental Politics* 11(Winter), 1–20.
Buck-Morss, Susan (1986). "The Flaneur, the Sandwichman and the Whore: the Politics of Loitering." *New German Critique* 39 (Autumn), 99–140.
Butler, Judith, and Joan W. Scott (Eds.) (1992). *Feminists Theorize the Political*. London: Routledge.
Calhoun, Craig (Ed.) (1992). *Habermas and the Public Sphere*. Cambridge, MA: MIT Press.
Canovan, Margaret (1998). "Hannah Arendt: Republicanism and Democracy." In April Carter and Geoffrey Stokes (Eds.). *Liberal Democracy and Its Critics*. Oxford, UK: Polity Press, pp. 39–57.
―― (1983). "A Case of Distorted Communication: A Note on Habermas and Arendt." *Political Theory* 11(Fall), 105–16.
Castoriadis, Cornelius (1997a). *The Castoriadis Reader*. Translated and Edited by David Ames Curtis. Oxford, UK: Blackwell Publishers.

Castoriadis, Cornelius (May 1997b). "Anthropology, Philosophy, Politics." *Thesis Eleven* 49, 99–116.
——— (1997c). "Democracy as Procedure and Democracy as Regime." *Constellations* 4(1), 1–18.
——— (1991). *Power, Politics and Autonomy*. Oxford, UK: Oxford University Press.
——— (1987). *The Imaginary Institution of Society*. Translated by Kathleen Blamey. Cambridge, MA: MIT Press.
Castronovo, Russ (2001). *Necro Citizenship: Death, Eroticism, and the Public Sphere*. Durham, NC, and London: Duke University Press.
Cheyfitz, Eric (1991). *The Poetics of Imperialism: Translation from Colonization from The Tempest to Tarzan*. Oxford, UK: Oxford University Press.
Christoff, Peter (1996). "Ecological Citizens and Ecologically Guided Democracy." In Brian Doharty and Marius de Geus (Eds.). *Democracy and Green Political Thought*. London: Routledge, pp. 151–69.
Christoff, Peter, John S. Dryzek, Robyn Eckersley, Robert E. Goodin, and Val Plumwood (2001). "Symposium: Green Thinking—from Australia." *Environmental Politics* 10(4), 85–102.
Code, Lorraine (2000). "How to Think Globally: Stretching the Limits of Imagination." In U. Narayan and S. Harding (Eds.). *Decentering the Centre*. Bloomington, IN: Indiana University Press, pp. 67–79.
——— (1995). *Rhetorical Spaces: Essays on Gendered Spaces*. London: Routledge.
——— (1993). "Taking Subjectivity into Account." In Linda Alcoff and Elizabeth Potter (Eds.). *Feminist Epistemologies*. London: Routledge, pp. 15–48.
——— (1991). *What Can She Know?: Feminist Theory and the Construction of Knowledge*. London: Cornell University Press.
Cohen, Joshua (1996). "Procedure and Substance in Deliberative Democracy." In James Bohman and William Rehg (Eds.). *Deliberative Democracy: Essays on Reason and Politics*. Cambridge, MA: MIT Press, pp. 407–38.
Cohen, Maurie (1996). "Risk Society and Ecological Modernisation: Alternative Visions for Postindustrial Nations." *OCEES Research Paper No.7*. Oxford, UK: Mansfield College.
Coles, Romand (August 2000). "Of Democracy, Discourse, and Dirt Value: Developments in Recent Critical Theory." *Political Theory* 28(4), 540–64.
Curry, Patrick (2000). "The Campaign for Political Ecology: Rethinking Green Politics, Book Review." *Environmental Values* 9(1), 120–22.
Curtin, Deane (1999). *Chinnagrounder's Challenge: The Question of Ecological Citizenship*. Indianapolis, IN: Indiana University Press.
Curtis, Kimberly (1999). *Our Sense of the Real: Aesthetic Experience and Arendtian Politics*. Ithaca, NY: Cornell University Press.
Darier, Éric (1999). "Foucault and the Environment: An Introduction." In Éric Darier (Ed.). *Discourses of the Environment*. Oxford, UK: Blackwell Publishers, pp. 1–34.

De Certeau, Michel (1984). *The Practice of Everyday Life*. Translated by Steven Rendall. Berkeley, CA: University of California Press.

Dean, Hartley (2001). "Green Citizenship." *Social Policy and Administration* 35(5), 490–505.

Debord, Guy (1998 [1988]). *Comments on the Society of the Spectacle*. Translated by Malcolm Imrie. London: Verso.

——— (1994 [1967]). *The Society of the Spectacle*. Translated by Donald Nicholson-Smith. New York: Zone Books.

D'Entrèves, Maurizio Passerin (1994). *The Political Philosophy of Hannah Arendt*. London: Routledge.

——— (1992). "Hannah Arendt and the Idea of Citizenship." In Chantal Mouffe (Ed.). *Dimensions of Radical Democracy: Pluralism, Citizenship, Community*. London: Verso, pp. 145–68.

Deveaux, Monique (1999). "Agonism and Pluralism." *Philosophy and Social Criticism* 25(4), 1–22.

Dietz, Mary (1996). "Context Is All: Feminism and Theories of Citizenship." In Chantal Mouffe (Ed.). *Dimensions of Radical Democracy: Pluralism, Citizenship, Community*. London: Verso, pp. 63–88.

Dobson, Andrew (2006). "Ecological Citizenship: A Defence." *Environmental Politics* 15(3), 447–51.

Dobson, Andrew and Angel Valencia Siaz (2005). "Introduction." *Environmental Politics* 14(2), 157–62.

——— (2005). "Globalization, Cosmopolitanism and the Environment." Accessed on June 7, 2005, available at http://www.open.ac.uk/social sciences/ccig/ccigsubset/cciginfopops/globalization_and_the_environment.pdf.

——— (2004). "Citizenship and the Ecological Challenge." Accessed on June 7, 2005, available at http://www.essex.ac.uk/ecpr/events/jointsessions/paperarchive/uppsala/ws5/Dobson.pdf.

——— (2003). *Citizenship and the Environment*. Oxford, UK: Oxford University press.

——— (1999). "Ecological Citizenship: A Disruptive Influence?" Accessed on June 25, 2002, available at www.psa.ac.uk/cps/1999/dobson.pdf.

Douthwaite, Julia (1997). "Homo Ferus: Between Monster and Model." *Eighteenth Century Life* 21(2), 176–202.

Dryzek, John (2001). "Legitimation and Economy in Deliberative Democracy." *Political Theory* 29, 651–69.

——— (2000). *Deliberative Democracy and Beyond: Liberals, Critics, Contestations*. Oxford, UK: Oxford University Press.

Eckersley, Robyn (1999). "The Discourse Ethic and the Problem of Representing Nature." *Environmental Politics* 8(2), 24–49.

——— (1990). "Habermas and Green Political Theory: Two Roads Diverging." *Theory and Society* 19, 739–76.

Eley, Geoff (1992). "Nations, Publics, and Political Cultures: Placing Habermas in the Nineteenth Century." In Craig Calhoun (Ed.). *Habermas and the Public Sphere*. Cambridge, MA: MIT Press, pp. 289–339.

Elliott, Anthony (2002). "The Social Imaginary: A Critical Assessment of Castoriadis's Psychoanalytic Theory." *American Imago* 59(2), 141–70.
Ely, John (August 1996). "Political, Civic and Territorial Views of Association." *Thesis Eleven* 46, 33–65.
Euben, J. Peter (1996). "Taking It to the Streets: Radical Democracy and Radicalizing Theory." In David Trend (Ed.). *Radical Democracy: Identity, Citizenship, and the State.* New York: Routledge, pp. 62–77.
Evernden, Neil (1992). *The Social Creation of Nature.* London: John Hopkins University Press.
——— (1985). *The Natural Alien.* Toronto: University of Toronto Press.
Feyerabend, Paul K. (1978). *Against Method: Outline of an Anarchist Theory of Knowledge.* London: Verso.
Fishkin, James S. (1995). *The Voice of People: Public Opinion and Democracy.* New Haven, CT: Yale University Press.
——— (1991). *Democracy and Deliberation: New Directions for Democratic Reform.* New Haven, CT: Yale University Press.
Fotopoulos, Takis (2001). "Our Aims." *Democracy & Nature* 7(1), 5–9.
——— (1997). *Towards an Inclusive Democracy: The Crisis of the Growth Economy and the Need for a New Liberatory Project.* London: Cassell.
Foucault, Michel (1988). *Politics, Philosophy, Culture: Interviews and Other Writings 1977–1984.* New York: Routledge.
Fraser, Nancy (1997). *Justice Interruptus: Critical Reflections on the "Postsocialist" Condition.* New York: Routledge.
Gabrielson, Teena (2008). "Green Citizenship: A Review and Critique." *Citizenship Studies* 12(4), 429–46.
Gabrielson, Teena, and Katelyn Parady (2010). "Corporeal Citizenship: Rethinking Green Citizenship through the Body." *Environmental Politics* 19(3), 374–91.
Gleber, Anke (1999). *The Art of Taking a Walk: Flânerie, Literature, and Film in Weimar Culture.* Princeton, NJ: Princeton University Press.
Goldman, Emma (1972). *Red Emma Speaks.* Compiled and Edited by Kates Shulman. New York: Vintage Books.
Goodin, Robert E. (1992). *Green Political Theory.* Cambridge, UK: Polity Press.
Guha, Ramachandra (Spring 1989). "Radical American Environmentalism and Wilderness Preservation: A Third World Critique." *Environmental Ethics* 11(1), 71–83.
Gülalp, Haldun (2006). *Citizenship and Ethnic Conflict: Challenging the Nation-State.* London: Routledge.
Gutman, Amy, and Denis Thompson (1996). *Democracy and Disagreement.* Cambridge, MA: Belknap Press of Harvard University Press.
Habermas, Jürgen (1999). "Popular Sovereignty as Procedure." In Stephen Macedo (Ed.). *Deliberative Politics: Essays on Democracy and Disagreement.* Oxford, UK: Oxford University Press, pp. 36–65.
——— (1998). *The Inclusion of the Other: Studies in Political Theory.* Cambridge, MA: MIT press.

――― (1996a). *Between Facts and Norms.* Cambridge, MA: MIT Press.
――― (1996b). "Three Models of Democracy." In Seyla Benhabib (Ed.). *Democracy and Difference: Contesting the Boundaries of the Political.* Princeton, NJ: Princeton University Press, pp. 21–30.
――― (1995). *Moral Consciousness and Communicative Action.* Cambridge, MA: MIT Press.
――― (1993). *Justification and Application: Remarks on Discourse Ethics.* Cambridge, MA: MIT Press, pp. 147–76.
――― (1990). *The Philosophical Discourse of Modernity: Twelve Lectures.* Cambridge, MA: MIT Press.
――― (1989). *The Structural Transformation of the Public Sphere: An Inquiry into a Category of Bourgeois Society.* Cambridge, MA: MIT Press.
――― (1987). *The Theory of Communicative Action Volume II: Lifeworld and System: A Critique of Functionalist Reason.* Cambridge, MA: MIT Press.
――― (1984). *The Theory of Communicative Action Volume I: Reason and the Rationalization of Society.* Cambridge, MA: MIT Press.
――― (1981). "The Dialectics of Rationalization: An Interview with Jürgen Habermas by Axel Honneth, Eberhard Knödler-Bunte and Arno Widmann." *Telos* 49(Fall), 5–31.
Hansen, Phillip (1993). *Hannah Arendt: Politics, History and Citizenship.* Stanford, CA: Stanford University Press.
Haraway, Donna (1991). *Simians, Cyborgs, and Women: The Reinvention of Nature.* New York: Routledge.
――― (1989). *Primate Visions: Gender, Race, and Nature in the World of Modern Science.* New York: Routledge.
Hayward, Bronwyn M. (1995). "The Greening of Participatory Democracy: A Reconstruction of Theory." *Environmental Politics* 4, 215–35.
Hayward, Tim (2006a). "Ecological Citizenship: Justice, Rights and the Virtue of Resourcefulness." *Environmental Politics* 15(3), 435–46.
――― (2006b). "Ecological Citizenship: A Rejoinder." *Environmental Politics* 15(3), 452–53.
――― (1998). *Political Theory and Ecological Values.* Cambridge, UK: Polity Press.
Heller, Chaia (1999). *Ecology of Everyday Life: Rethinking the Desire for Nature.* Montreal: Black Rose Books.
Herzog, Annabel (2004). "Political Itineraries and Anarchic Cosmopolitanism in the Thought of Hannah Arendt." *Inquiry* 47, 20–41.
Heyd, Thomas (2003). "Bashō and the Aesthetics of Wandering: Recuperating Space, Recognising Place, and Following the Ways of the Universe." *Philosophy East and West* 53, 291–307.
Hicks, Darrin (2002). "The Promise(s) of Deliberative Democracy." *Rhetoric & Public Affairs* 5(2), 223–60.
Honig, Bonnie (2001). *Democracy and the Foreigner.* Princeton, NJ: Princeton University Press.
――― (1994). "Difference, Dilemmas and the Politics of Home." *Social Research* 61(3), 563–97.

Honig, Bonnie (1992). "Toward an Agonistic Feminism: Hannah Arendt and the Politics of Identity." In Judith Butler and Joan W. Scott (Eds.). *Feminists Theorize the Political*. New York: Routledge, pp. 215–35.
——— (1991). "Declarations of Independence: Arendt and Derrida on the Problem of Founding a Republic." *American Political Science Review* 85(Winter), 97–113.
hooks, bell (1984). *Feminist Theory: From Margin to Centre* Boston, MA: South Wend Press.
Horowitz, Irving Louis (1999). "Totalitarian Visions of the Good Society: Arendt." *Partisan Review* 66, 263–79.
Hyde, Lewis (1998). *Trickster Makes This World: Mischief, Myth, and Art*. New York: North Point Press.
Isaac, Jeffrey (1998). *Democracy in Dark Times*. Ithaca, NY: Cornell University Press.
——— (1994). "Oasis in the Desert: Hannah Arendt on Democratic Politics." *American Political Science Review* 88(1), 156–68.
Itard, Jean-Marc-Gaspard (1962). *The Wild Boy of Aveyron*. New York: Appleton-Century-Crofts.
Ivie, Robert (2002). "Rhetorical Deliberation and Democratic Politics in the Here and Now." *Rhetoric and Public Affairs* 5(2), 277–85.
——— (1998). "Democratic Deliberation in a Rhetorical Republic." *Quarterly Journal of Speech* 84, 491–505.
Jameson, Frederic (1991). *Postmodernism or, the Cultural Logic of Late Capitalism*. Durham, NC: Duke University Press.
Jappe, Anselm (1999). *Guy Debord*. Translated by Donald Nicholson-Smith. Berkeley, CA: University of California Press.
Jay, Martin (1978). "Hannah Arendt: Opposing Views." *Partisan Review* 45, 348–80.
Jelin, Elizabeth (2000). "Towards a Global Environmental Citizenship?" *Citizenship Studies* 4(1), 47–63.
Johnson, James, and Dana Villa (1994). "Public Sphere, Postmodernism and Polemic." *American Political Science Review* 88 (June), 427–33.
Kalyvas, Andreas (2004). "From the Act to the Decision: Hannah Arendt and the Question of Decisionism." *Political Theory* 32, 320–46.
——— (2001). "The Politics of Autonomy and the Challenge of Deliberation: Castoriadis Contra Habermas." *Thesis Eleven* 64, 1–19.
——— (1999). "Reviews." *Thesis Eleven* 59, 103–11.
Kapoor, Ilan (2004). "Hyper-Self-Reflexive Development? Spivak on Representing the Third World 'Other.'" *Third World Quarterly* 25(4), 627–47.
———(2003). "Acting in a Tight Spot: Homi Bhabha's Postcolonial Politics." *New Political Science* 25(4), 561–77.
Karagiannis, N, and P. Wagner (2005) "Towards a Theory of Synagonism." *Journal of Political Philosophy* 13(3), 235–262.
Kassiola, Joel Jay, and Sujian Guo (Eds.) (2010). *China's Environmental Crisis: Domestic and Global Political Impacts and Responses*. New York: Palgrave Macmillan.

Keenan, Alan (2003). *Democracy in Question: Democratic Openness in a Time of Political Closure*. Stanford, CA: Stanford University Press.

Kharkhordin, Oleg (2001). "Nation, Nature and Natality: New Dimensions." *European Journal of Social Theory* 4(4), 459–78.

Knabb, Ken (1997). *Public Secrets: Collective Skirmishes of Ken Knabb: 1970–97.* Berkeley, CA: Bureau of Public Secrets.

——— (1981). *Situationist International: Anthology*. Berkeley, CA: Bureau of Public Secrets.

Knight, Jack, and James Johnson (1997). "What Sort of Equality Does Deliberative Democracy Require?" In James Bohman and William Rehg (Eds.). *Deliberative Democracy: Essays on Reason and Politics*. Cambridge, MA: MIT Press, pp. 279–319.

Kulynych, Jessica (1997). "Performing Politics: Foucault, Habermas, and Postmodern Participation." *Polity* 30(Winter), 315–46.

Kurasawa, Fuyuki (2000). "At the Crossroads of the Radical: The Challenges of Castoriadis's Thought." *Theory, Culture & Society* 17(4), 145–55.

Laclau, Ernesto (1990). *New Reflections on the Revolution of Our Time*. London: Verso.

Laclau, Ernesto, and Chantal Mouffe (1985). *Hegemony and Socialist Strategy: Towards a Radical Democratic Politics*. London: Verso.

Latour, Bruno (2004). *Politics of Nature: How to Bring the Science into Democracy*. Translated by Catherine Porter. London: Harvard University Press.

Latta, P. Alex (2007). "Locating Democratic Politics in Ecological Citizenship." *Environmental Politics* 16(3), 377–93.

Leca, Jean (1992). "Questions on Citizenship." In Chantal Mouffe (Ed.). *Dimensions of Radical Democracy: Pluralism, Citizenship, Community*. London: Verso, pp. 17–32.

Leet, Martin (1998). "Jürgen Habermas and Deliberative Democracy." In A. Carter and G. Stokes (Ed.). *Liberal Democracy and Its Critics*. Cambridge, UK: Polity Press, pp. 77–97.

Leiber, Justin (1997). "Nature's Experiments, Society's Closures." *Journal for the Theory of Social Behaviour* 27(2/3), 325–43.

Lefort, Claude (1988). *Democracy and Political Theory*. Minneapolis, MN: University of Minnesota Press.

Liu, Yameng (1999). "Justifying My Position in Your Terms: Cross-Cultural Argumentation in a Globalized World." *Argumentation* 13, 297–315.

Lorde, Audre (1997). "Age, Race, Class, and Sex: Women Redefining Difference." In Anne McClintock, Aamir Mufti, and Ella Shohat (Eds.). *Dangerous Liaisons: Gender, Nation, & Postcolonial Perspectives*. Minnesota, MN: University of Minnesota Press, pp. 374–80.

Lowe, Carmen (Fall 2000). "Where the Country of Lost Borders Meets Jeffers Country: The Walking Women of Robinson Jeffers and Mary Austin." *Jeffers Studies* 4(4), 21–46.

Macedo, Stephen (Ed.) (1999). *Deliberative Politics: Essays on Democracy and Disagreement*. Oxford, UK: Oxford University Press.

MacGregor, Sherilyn (2006). "No Sustainability without Justice: A feminist Critique of Environmental Citizenship." In Andrew Dobson and Derek Bell (Eds.). *Environmental Citizenship*. Cambridge: MIT Press, pp. 101–27.
——— (2004). "From Care to Citizenship: Calling Ecofeminism Back to Politics." *Ethics and the Environment* 9(1), 56–84.
——— (2001). "Beyond Mothering Earth: Ecological Citizenship and the Gendered Politics of Care." Unpublished Dissertation. North York: York University.
Maclean, Charles (1977). *The Wolf Children*. London: Allen Lane.
Malatesta, Errico (1965). *Errico Malatesta: His Life and Ideas*. Compiled and Edited by Vernon Richards. London: Freedom Press.
Malson, Lucien (1972). *Wolf Children*. Translated in 1964 by Edmund Fawcett, Peter Ayerton, and Joan White. London: NLB.
Mansbridge, Jane (1999). "Everyday Talk in the Deliberative System." In Stephen Macedo (Ed.). *Deliberative Politics: Essays on Democracy and Disagreement*. Oxford, UK: Oxford University Press, pp. 211–39.
——— (1996). "Using Power/Fighting Power: The Polity." In Seyla Benhabib (Ed.). *Democracy and Difference: Contesting the Boundaries of the Politica*. Princeton, NJ: Princeton University Press, pp. 46–66.
Markell, Patchen (1997). "Contesting Consensus: Rereading Habermas on the Public Sphere." *Constellations* 3(3), 377–400.
Marotta, Vince. (2000). "The Stranger and Social Theory." *Thesis Eleven* 62, 121–34.
Mason, Paul (2012). *Why It's Kicking off Everywhere: The New Global Revolutions*. London: Verso.
McClintock, Anne, Aamir Mufti, and Ella Shohat (Eds.) (1997). *Dangerous Liaisons: Gender, Nation, & Postcolonial Perspectives*. Minneapolis, MN: University of Minnesota Press.
Mendoza, Lily S. (2001). *Between the Homeland and Diaspora: The Politics of Theorizing Filipino and Filipino American Identities: A Second Look at the Poststructuralism-Indigenization Debate*. New York: Routledge.
Mignolo, Walter D. (2002). "The Many Faces of Cosmo-polis: Border Thinking and Critical Cosmopolitanism." In Carol A. Breckenridge, Sheldon Pollock, Homi K Bhabha, and Dipesh Chakrabarty (Eds.). *Cosmopolitanism*. Durham, NC: Duke University Press, pp. 157–87.
——— (2000). "The Many Faces of Cosmo-polis: Border Thinking and Critical Cosmopolitanism." *Public Culture* 12(3), 721–48.
Mouffe, Chantal (2005). *On the Political*. London: Routledge.
——— (2000). *The Democratic Paradox*. London: Verso.
——— (1999). "Deliberative Democracy or Agonistic Pluralism." *Social Research* 66(3), 745–58.
——— (1995). "Post-Marxism: Democracy and Identity." *Environment and Planning; Society and Space*. 13, 259–265.
——— (1993). *The Return of the Political*. London: Verso.
——— (1992a). *Dimensions of Radical Democracy: Pluralism, Citizenship, Community*. London: Verso.

——— (1992b). "Democratic Citizenship and the Political Community." In *Dimensions of Radical Democracy: Pluralism, Citizenship, Community*. London: Verso, pp. 225–39.

Newby, Howard (1996). "Citizenship in a Green World: Global Commons and Human Stewardship." In M. Bulmer and A. Rees (Eds.). *Citizenship Today: The Contemporary Relevance of T. H. Marshall*. London: UCL Press, pp. 209–21.

Newton, Michael (2002). *Savage Girls and Wild Boys: A History of Feral Children*. London: Faber and Faber.

Noske, Barbara (1997). *Beyond Boundaries: Humans and Animals*. Montreal: Black Rose Books.

Oikonomou, Yorgos (July 2003). "Plato and Castoriadis: The Concealment and the Unveiling of Democracy." *Democracy and Nature* 9(2), 247–62.

Palti, Elias Jose (1998). "Reviews." *Thesis Eleven* 54(August), 117–22.

Pels, Dick (1999). "Privileged Nomads: On the Strangeness of Intellectuals and the Intellectuality of Strangers." *Theory, Culture and Society* 16(February), 63–86.

Pitkin, Hanna (1981). "Justice: On Relating Private and Public." *Political Theory* 9(1), 327–52.

Plumwood, Val (2002). *Environmental Culture: The Ecological Crisis of Reason*. New York: Routledge.

Pollock, Sheldon, Homi K. Bhabha, Carol A. Breckenridge, and Dipesh Chakrabarty (2000). "Cosmopolitanisms." *Public Culture* 12(3), 577–89.

Reid, Herbert, and Betsy Taylor (2000). "Embodying Ecological Citizenship: Rethinking the Politics of Grassroots Globalization in the United States." *Alternatives* 25, 439–66.

Sachs, Wolfgang (1999). *Planet Dialectics: Explorations in Environment and Development*. New York: Zed Books.

Said, Edward W. (2000). *Reflections on Exile and Other Essays*. Cambridge, MA: Harvard University Press.

——— (1994). *Representations of the Intellectual: The 1993 Reith Lectures*. New York: Vintage Books.

——— (1990). "Reflections on Exile." In Russel Ferguson (Ed.). *Out There: Marginalization and Contemporary Culture*. Cambridge, MA: MIT Press.

Sanders, Lynn M (1997). "Against Deliberation." *Political Theory* 25(3), 347–76.

Sandilands, Catriona (2003). "A Flâneur in the Forest? Strolling Point Pelee with Walter Benjamin." *Topia* 3, 37–57.

——— (2002). "Opinionated Natures: Toward a Green Public Culture." In Bob Pepperman Taylor and Ben Minter (Eds.). *Democracy and the Claims of Nature: Critical perspectives for a New Century*. Lanham, MD: Rowan and Littlefield, pp. 117–32.

——— (2000). "Raising Your Hand in the Council of All Beings: Ecofeminism and Citizenship." *Ethics and the Environment* 4(6), 219–33.

——— (1999). *The Good Natured Feminist: Ecofeminism and the Quest for Democracy*. Minneapolis, MN: University of Minnesota Press.

Saward, Michael (2003). *Democracy*. Oxford, UK: Blackwell Publishing.
——— (1993). "Green Democracy?" In A. Dobson and P. Lucardie (Eds.). *The Politics of Nature: Explorations in Green Political Theory*. London: Routledge, pp. 63–80.
Sen, Amartya (2000). "Other People: Beyond Identity." *New Republic* 18(December), 26–27.
——— (1999). *Development as Freedom*. New York: Random House.
Shafir, Gershon (Ed.) (1998). *The Citizenship Debates: A Reader*. Minneapolis, MN: University of Minnesota Press.
Shattuck, Roger (1996). *Forbidden Kowledge*. New York: St. Martin Press
——— (1980). *The Forbidden Experiment: The Story of the Wild Boy of Aveyron*. London: Quartet Books.
Singh, J. A. L., and Robert Zingg (1942). *Wolf-Children and Feral Man*. New York: Harpers.
Smith, Mark (1998). *Ecologism: Towards Ecological Citizenship*. Milton Keynes: Open University Press.
Solnit, Rebecca (2000). *Wanderlust, A History of Walking*. New York: Viking Press.
Spivak, Gayatri (1996). In D. Landry and G. MacLean (Eds.). *The Spivak Reader*. New York: Routledge.
——— (1988). "Can the Subaltern Speak?" In C. Nelson and L. Grossberg (Eds.). *Marxism and Interpretation of Culture*. Chicago, IL: University of Illinois Press, pp. 271–313.
Steeves, Peter H. (2003). "The Familiar Other and Feral Selves: Life at the Human/Animal Boundary." In Angela N. H. Creager and William Chester Jordan (Eds.). *The Animal/Human Boundary: Historical Perspectives*. Rochester, NY: University of Rochester Press, pp. 228–64.
Stevenson, Nick (September 2003). "Cultural Citizenship in the 'Cultural' Society: A Cosmopolitan Approach." *Citizenship Studies* 7(3), 331–48.
Steward, Fred (1991). "Citizens of Planet Earth." In G. Andrews (Ed.). *Citizenship*. London: Lawrence and Wishart, pp. 65–75.
Stichweh, Rudolf (1997). "The Stranger—On the Sociology of Indifference." *Thesis Eleven* 51, 1–16.
Taussig, Michael (1987). *Shamanism, Colonialism and the Wildman*. Chicago: University of Chicago Press.
Tester, Keith (Ed.) (1994). *The Flâneur*. New York: Routledge.
Thoreau, Henry David (1992). *Walden and Resistance to Civil Government*. New York: Norton.
——— (1993). "Walking." In Susan J. Armstrong and Richard G. Botzler (Eds.). *Environmental Ethics: Divergence and Convergence*. New York: McGraw-Hill, pp. 108–17.
Torfing, Jacob (1999). *New Theories of Discourse: Laclau, Mouffe and Žižek*. Oxford, UK: Blackwell Publishers.
Torgerson, Douglas (2000). "Farewell to the Green Movement? Political Action and the Green Public Sphere." *Environmental Politics* 9(4), 1–19.

—— (1999). *The Promise of Green Politics: Environmentalism and the Public Sphere.* Durham, NC: Duke University Press.

Turner, Bryan (1992). "Outline of a Theory of Citizenship." In Chantal Mouffe (Ed.). *Dimensions of Radical Democracy: Pluralism, Citizenship, Community.* London: Verso, pp. 33–62.

Urry, John (1998). "Globalization and Citizenship" (draft). Lancaster: Department of Sociology. Accessed on June 17, 2002, available at http://www.comp.lancaster.ac.uk/sociology/soc009ju.html.

Valadez, Jorge (2001). *Deliberative Democracy, Political Legitimacy, and Self-Determination in Multicultural Societies.* Boulder, CO: Westview Press.

Van Steenbergen, Bart (1994). "Toward a Global Ecological Citizen." In Bart Van Steenbergen (Ed.). *The Condition of Citizenship.* London: Sage.

Vicas, Astrid (1998). "Reusing Cultures: The Import of Detournement." *The Yale Journal of Criticism* 11(2), 381–406.

Villa, Dana R. (1998). "The Philosopher versus the Citizen: Arendt, Strauss, and Socrates." *Political Theory* 26(2), 147–72.

—— (1992). "Postmodernism and the Public Sphere." *American Political Science Review* 86(September), 712–21.

Wallace, Anne D. (1993). *Walking, Literature, and English Culture: The Origins and Uses of Peripatetic in the Nineteenth Century.* Oxford, UK: Clarendon Press.

Walzer, Michael (1999). "Deliberation and What Else?" In Stephen Macedo (Ed.). *Deliberative Politics: Essays on Democracy and Disagreement.* New York: Oxford University Press, pp. 58–69.

—— (1994). "The Civil Society Argument." In Gershon Shafir (Ed.). *The Citizenship Debates: A Reader.* Minneapolis. MN: University of Minnesota Press, pp. 291–308.

Warner, Michael (2002). "Publics and Counterpublics." *Public Culture* 14(1), 49–90.

—— (1992). "The Mass Public and the Mass Subject." In Calhoun Craig (Ed.). *Habermas and the Public Sphere.* Cambridge, MA: MIT Press, pp. 377–401.

White, Heyden (1972). "The Forms of Wildness: Archaeology of an Idea." In Edward Dudley and Maximillion Novak (Eds.). *The Wild Man Within.* Pittsburgh, PA: University of Pittsburgh Press, pp. 150–82.

Whitworth, Andrew (2001). "Ethics and Reality in Environmental Discourses." *Environmental Politics* 10(2), 22–42.

Wolff, Janet (1985). "The Invisible Flâneuse: Women and the Literature of Modernity." *Theory, Culture & Society* 2, 37–26.

Wolin, Sheldon (1996). "Fugitive Democracy." In Seyla Benhabib (Ed.). *Democracy and Difference: Contesting the Boundaries of the Political.* Princeton, NJ: Princeton University Press, pp. 31–45.

—— (1992). "What Revolutionary Action Means Today." In Chantal Mouffe (Ed.). *Dimensions of Radical Democracy: Pluralism, Citizenship, Community.* London: Verso, pp. 240–53.

Wolin, Sheldon (1983). "Hannah Arendt: Democracy and the Political." *Salmagundi* 60(Spring/Summer), 3–19.
Young, Iris Marion (2001). "Activist Challenges to Deliberative Democracy." *Political Theory* 29, 670–90.
—— (2000). *Inclusion and Democracy.* Oxford, UK: Oxford University Press.
—— (1997a). "Difference as a Resource for Democratic Communication." In James Bohman and William Rehg (Eds.). *Deliberative Democracy: Essays on Reason and Politics.* Minneapolis, MN: University of Minnesota Press, pp. 383–406.
—— (1997b). "Asymmetrical Reciprocity: On Moral Respect, Wonder, and Enlarged Thought." *Constellations* 3(3), 340–63.
—— (1996). "Communication and the Other: Beyond Deliberative Democracy." In Seyla Benhabib (Ed.). *Democracy and Difference: Contesting the Boundaries of the Political.* Princeton, NJ: Princeton University Press, pp. 120–35.
—— (1994). "Polity and Group Difference: A Critique of the Ideal of Universal Citizenship." In Gershon Sharif (Ed.). *The Citizenship Debates: A Reader.* Minneapolis, MN: University of Minnesota Press, pp. 263–91.
Yousef, Nancy (2001). "Savage or Solitary?: The Wild Child and Rousseau's Man of Nature." *Journal of the History of Ideas* 62(2), 245–63.
Zerilli, Linda (1998). "This Universalism Which Is Not One." *Diacretics* 28(2), 3–20.
Žižek, Slavoj (1990). "Beyond Discourse-Analysis." In Ernesto Laclau (Ed.). *New Reflections on the Revolution of Our Time.* London: Verso, pp. 249–59.
—— (1989). *The Sublime Object of Ideology.* London: Verso.

INDEX

Abbey, Edward 129
active listening 66, 69–71, 73, 130, 150–1
agency *see* political
agonism 86, 95, 98–9, 120
agonistic politics 88, 95, 97–9, 113
Agyerman, Julian and Bob Evans 142
Animal Rights Canada 47
antagonism 2, 10, 65, 86, 88, 95–9, 103, 126
  see also agonistic politics
anthropocentrism 45, 47, 51, 125–6, 137
antiausterity 87–92
antiauthoritarian 15, 58, 82, 146, 152
antiessentialism 57, 113
Anzaldúa, Gloria 67–9
Arab Spring 92
Arblaster, Anthony 6, 56, 146
Arendt, Hannah 1, 4, 17, 32, 41, 57, 58, 59–61, 65–7, 70, 75, 78–82, 87–8, 92, 96–8, 100–3, 107, 112–15, 122, 131, 134, 136, 149, 151, 153–4
Arias-Maldonado, Manuel 128
Austin, Mary 29–33
autonomous
  individuals 53, 68, 98–102, 154
  society 102, 145, 154

Barber, Benjamin 2
Barry, John 105, 124–6, 142
Barry, Wendell 129
Baudelaire, Charles 22–4
Bauman, Zigmund 26

Benjamin, Walter 23, 26–7
Bhabha, Homi 67–8, 81
Bickford, Susan 55, 68, 70, 72, 81
Birch, Thomas 18
Bookchin, Murray 11, 106, 141
borders 19, 41, 45–6, 78, 81, 113, 122
border thinking 67–8, 75
Braidotti, Rosi 72–5
Buck-Morss, Susan 27

care ethic 33, 124, 128–32
Castoriadis, Cornelius 3, 87–8, 98–100, 145–6, 153
Christoff, Peter 124–6
Code, Lorraine 77–8
commodification 26
commodity 35, 37, 39
common good 63, 108, 110–11
communicative
  action 132, 134, 152
  communities 138
  discourse 136, 139
  ethics 136, 137, 139
  moment 130–1
  power 93
consensus 86–7, 93–4, 96, 97, 99, 107, 136, 137, 153
cosmopolitan 23, 118–20, 125
Curtin, Deane 128–31, 133

D'Entrèves, Maurizio Passerin 59, 60
De Certeau, Michel 20
Dean, Hartley 109, 110, 128, 135–6
Debord, Guy 14, 35–40
decision making 14–16, 92, 94–5, 110, 123–4, 137–40, 155

Index

democracy
  deliberative 53, 95, 135–7, 140, 153, 155
  liberal 3, 41, 85–8, 97, 102, 136, 142, 150, 153
  plural 87, 95
  radical 5, 85–6, 95, 97
democratic
  creation 3, 100
  ethics 3, 9, 17, 34, 58, 83, 117, 124, 129, 131, 136, 140, 143, 145–7, 149
  terrain 6, 9, 17, 25, 42, 56–8, 62–3, 65, 145, 146, 148, 149, 151
dérive 38–9
desiring subject 66–7, 69
détournement 38
disruption 3–4, 7–9, 11, 21, 36, 38, 43, 45–8, 52, 55, 63, 70–2, 75–7, 80–3, 91, 98, 103, 106, 108, 112, 116–18, 121, 123, 141, 145–8, 150, 154
dissent 102, 145
Dobson, Andrew 107, 117–21, 128
Dryzek, John 136–9

Eckersley, Robyn 105, 136–8
ecocentric desire 137
ecocommunicative 136, 137
ecocommunitarian 128
ecodeliberative ethics 132, 137–8
ecological citizenship *see* green citizenship
ecological contract 125
ecological ethic 124
ecological footprint 118, 120
enviro-doxa 111
environmental
  activism 134
  desires 4
  movement 11, 85, 103–4, 123, 154
  politics 10, 105–7, 110, 118, 122, 124
  thought 105, 109, 129

environmentalism 5, 11, 103–4, 106, 110, 122–4, 133, 140, 142, 154–5
environmentalism-ecology partnership 109, 111
environmentalists 4, 46, 85, 104, 108, 111–12, 115–16, 120, 123, 128, 132, 140, 142–3, 154, 156
epistemology 77–8
ethic of care 128, 130

Fanon, Frantz 66–9
feminism
  difference 71–6, 82
  sisterhood 71, 74, 96
feral animals 45–8, 54
feral children 45–6, 48–51
flâneur 22–6
flâneuse 26–9
forbidden Britain 18
forbidden experiment 48, 52–3
forces of Freedom 1–3, 12, 72, 109, 126, 150
forgiveness 4, 62
Fotopoulos, Takis 1
Foucault, Michel 7
friendly enemy 11, 99

Gabrielson, Teena 108–9, 142
gender 26–34, 76
genuine questions 100
Gleber, Anke 23, 26, 27
Goldman, Emma 149, 152
Goodin, Robert 110
green citizenship 103–7, 109, 114–16, 121–6, 128, 142
green public sphere 109, 122
Guys, Constantin 22–4

Habermas, Jürgen 86–95, 139, 152, 154
Haraway, Donna 75–8
Honig, Bonnie 113–14
Hooks, bell 71–2, 96, 151

INDEX

imaginary 3, 100, 132
  radical 98
  social 50, 98
instituting-insituted society 97–100, 104, 149
intersubjective communication 67, 137
ISS (Ideal Speech Situation) 69, 86, 136–8
Itard Jean-Marc-Gaspard 50–1
Ivie, Robert 145

Jappe, Anselm 38
Jefferson, Thomas 129
Jelin, Elizabeth 105, 110

Keenan, Alan 154
Kinder Scout 19
Knabb, Ken 37–9
Kulynych, Jessica 101

Laclau, Ernesto 87
  see also Mouffe, Chantal
law 18–19, 68, 89, 93–4, 106, 154
legislation 89, 91, 93–4, 103
legitimacy 103, 104, 138, 140, 145, 146, 154–6
Leiber, Joseph 49
liberal democracy 2–3, 17, 41, 74, 85–8, 93, 97, 102, 125, 136, 138, 149, 150, 153
listening *see* active listening
Lorde, Audre 72
Lowe, Carmen 30–2

MacGregor, Sherilyn 142
Madame Guérin 50
Madame Recamier 49
making face 68
Malatesta Errico 105
marches 8, 40–1
Mason, Paul 61
media 61–2, 92, 98, 146, 150
Mestiza consciousness 67–8
micropolitical moments 9, 59, 82, 147

Missouri Department of Conservation 46–7
Mouffe, Chantal 3–4, 57, 88, 110, 112, 123
  and Ernesto Laclau 5, 86–7, 150

natality 98, 127
nature 30, 48
needy 45, 130
politicization of 11, 106, 114, 123, 143, 154–5
pure state of 48, 52–3
representation of 106, 111, 114–15, 132, 136–40, 155
neoliberalism 53
Newton, Michael 49, 51, 53–4
Nietzsche, Friedrich 42
noble savage 49, 53
nomadic consciousness 73–5
nomads 7, 55, 74,
Noske, Barbara 46

obligations 75, 117, 121, 127
Occupy movement 15, 61, 87, 91–2
otherness 67, 114, 116, 131, 134

Painter of Modern Life, The 22–4
Palti, Elias Jose 94
pedestrian 20–1, 23
performance 42, 60–2, 78, 80–2, 92, 96, 112, 143, 152
peripatetic 25, 34
pilgrimage 8, 14, 40–1
Pitkin, Hannah 114–15
Plumwood, Val 138–9, 156
pluralist society 86, 95, 111, 143
plurality 56–7, 59, 61, 69, 79, 97, 98–101, 108, 112–13, 123, 135, 148, 154
political
  agency 156
  dissent 145
  life 101
  moments 58, 61, 146, 156

political sphere *see* public sphere
politics 14–16, 97, 98, 100
postcolonial 82, 150
primacy of the political 56, 81, 95, 155–6
public good 111
public space 1, 10, 18, 56, 60, 75, 85, 88, 101, 116, 139
public sphere
  green 11, 106, 109–10, 122–3, 141–52, 155
  strong 98–102, 152
  weak 89–91, 94, 152–3

ramblers 8, 18–20, 28,
reciprocity 69, 72, 75
Reclaim the Streets 41
Reid, Herbert and Betsy Taylor 132–5
representative thinking 4, 59–60
republican 4, 87–8, 98, 113
responsibilities 32, 34, 56, 61, 75, 85, 89
  environmental 116, 120–1, 125, 147
revolution 7, 11, 37, 71, 99–100, 149
  democratic 5, 86–7, 110, 150
Right to Roam 18–20
Rousseau, Jean Jacques 49, 53

Sandilands, Catriona 112
Sands, George 27–9
Saward, Michael 2
Sen, Amartya 1, 8, 34
situated knowledge 76–8
Situationists 37–9
Smith, Mark 127
social movements 13–14, 40, 60–2, 85

Solnit, Rebecca 6, 15, 20, 40, 116
spectacle 35–7, 60
state, the 60–1, 91–2, 124, 146
Steeves, Peter 50–4
Steward, Fred 128
stewardship Ethic 115, 124–7
storyteller 60, 148
storytelling 132–5
survival 127
  spirit 115
synagonism 82–3, 146

Tester, Keith 25
theater 59, 61–2, 80–1, 151–2
Thoreau, Henry David 16–17, 90
Torgerson, Douglas 106–7, 122–3
treadmill 14–15
trespassing 19–20, 27, 28, 43
typical dynamic 14, 40, 62, 146

Van Steenbergen, Bart 107
Victor of Aveyron 48–51, 52
Villa, Dana 86
visiting 59–60
voting 14–16, 21

walking 16, 20, 132–5
Walking Woman, The 29–33
wanderers 21–2, 58
wandering 6–7, 13, 16–17, 28–9, 41, 62–3, 74, 116, 134, 146
"We" spaces 9–10, 65, 67, 72, 73, 75, 79, 96
Wolff, Janet 27

Young, Iris Marion 69, 71, 154
Yousef, Nancy 50–1

GE
180
G37
2013

Reinsch Library
Marymount University
2807 N Glebe Road
Arlington, VA 22207

Printed in the United States of America